John Tulloch

HIGH FREQUENCY MEASUREMENTS AND NOISE IN ELECTRONIC CIRCUITS

Douglas C. Smith

VAN NOSTRAND REINHOLD
New York

Library of Congress Catalog Card Number 92-4845
ISBN 0-442-00636-5

I(T)P Van Nostrand Reinhold is an International Thomson Publishing company.
ITP logo is a trademark under license.

Printed in the United States of America

Van Nostrand Reinhold   International Thomson Publishing GmbH
115 Fifth Avenue   Königswinterer Str. 418
New York, NY 10003   53227 Bonn
                                      Germany

International Thomson Publishing   International Thomson Publishing Asia
Berkshire House,168-173   221 Henderson Bldg. #05-1
High Holborn, London WC1V 7AA   Singapore 0315
England

Thomas Nelson Australia   International Thomson Publishing Japan
102 Dodds Street   Kyowa Building, 3F
South Melbourne 3205   2-2-1 Hirakawacho
Victoria, Australia   Chiyoda-ku, Tokyo 102
                                      Japan

Nelson Canada
1120 Birchmount Road
Scarborough, Ontario
M1K 5G4, Canada

16 15 14 13 12 11 10 9 8 7 6 5 4 3 2

Library of Congress Cataloging-in-Publication Data

Smith, Douglas C. (Douglas Charles), 1947-
    High frequency measurements and noise in electronic circuits: a
practical guide of successful techniques for designing, debugging, and
reducing noise / Douglas C. Smith.
        p. cm.
    Includes index.
    ISBN 0-442-00636-5
    1. Electronic circuits—Noise. 2. Electronic circuit design.
3. Electromagnetic noise. I. Title.
TK7867.5.sbb   1992
621.3815'43—dc20

To my loving wife Deborah and our kids,

Doug, Didi, and David,

and also my friends and associates for their help

especially George A. Florio,

Professor L. Ensign Johnson of Vanderbilt University,

W. Michael King, and Henry W. Ott

for providing inspiration and help not only

for this book but throughout my career.

Contents

Chapter 7 — Magnetic Loop and Other Noncontact Measurements

Preface

Necessity is said to be the mother of invention. Over the years, I have developed a philosophy and a set of methods for the purpose of measuring high frequency signals and noise in electronic circuits and for characterizing and mitigating high frequency noise (such as that caused by digital logic or switching power supplies) in electronic circuits as well. This development was made necessary by the realization that common measurement methods, such as with the common passive high impedance scope probe and oscilloscope, could yield errors of 100 percent and more in the amplitude of a waveform. Often I noticed that the process of trying to measure a signal or noise in a circuit affected its operation, further diminishing my confidence in a measurement.

In this book, I discuss a philosophy of measurement that ensures accurate results. In addition, new measurement techniques are presented later in the book that can yield accurate measurement of signals or noise in a circuit, and permit the characterization and location of sources of noise that heretofore were difficult if not impossible to deal with.

After reading this book, the reader should be able to measure signals and noise in electronic circuits as well as locate and reduce, if necessary, noise generated by electronic circuits or external interference. Most, hopefully all, of the "magic" of dealing with topics like ground noise, ground loops, location of noise sources, and other ElectroMagnetic Interference (EMI) effects will be removed and the reader will be able to characterize and mitigate noise in circuits with confidence.

This book is intended for practicing electrical engineers and students of electrical engineering at the undergraduate level. Readers are welcome to contact me with questions or comments at P. O. Box 524, Rumson, NJ 07760-0524.

HIGH FREQUENCY MEASUREMENTS AND NOISE IN ELECTRONIC CIRCUITS

1

Introduction

For many engineers and students, high frequency design and measurement techniques contain a liberal dose of "magic." This is unfortunate as these techniques can be explained in reasonably simple terms that allow their use in day-to-day engineering. The purpose of this book is to develop a measurement philosophy and a set of techniques that will allow the reader to measure high frequency signals and noise for the purpose of verifying design intent and to debug the design if problems are found. The techniques presented in this book are useful for measurements ranging from the noise margin in a digital circuit, to tracking down the effect of switching power supply noise on system operation, and even to evaluating system response to an Electrostatic Discharge (ESD) event.

In the process of investigating measurements, some discussion of noise mitigation techniques is difficult to avoid. These additional techniques will come as "icing on the cake" for the reader. After reading this book, the reader should be able to approach high frequency design, measurement, and noise mitigation techniques with confidence and realize that the subject can even be fun, not withstanding that sometimes measurements of this type are made late at night in a crisis atmosphere.

This is a good place to ask the question: What is meant by "high frequency" in the context of measurement and design? The working definition to be used in this book is the following: High frequency is that frequency above which parameters of the circuit that are normally considered parasitic and to be ignored, such as the inductance of a few inches of wire, become important enough to affect circuit operation. This can happen anywhere from a few MHz to several GHz depending on the design. The frequency range from 20 to 500 MHz is critical to a wide range of electronic equipment and is the primary frequency range that will be discussed in this book. Extension of the principles discussed to higher frequencies, in the GHz range, will be made where appropriate.

MEASUREMENT PHILOSOPHY AND TOPICS COVERED

This book is as much about measurement philosophy as any particular measurement technique. Throughout the book, a philosophy of measurement will be developed. It is hoped that by discussing several important measurement and noise mitigation techniques, the reader will be able to extend the principles and underlying philosophy to any situation requiring accurate measurements. Typical of the kind of questions the reader should find answers to are:

1. Why does connection of an oscilloscope to a malfunctioning circuit sometimes cause the circuit to work properly? This can be a frustrating experience, enough so as to tempt a design engineer to leave the scope probe in the circuit!
2. What factors can cause the result of a measurement to conflict with observed circuit observation? Occasionally, signal distortion or noise levels are measured that would seem to preclude proper circuit operation, yet the circuit operates perfectly! Conversely, the circuit does not operate, yet no reason can be uncovered.
3. How can noise be measured if it can not be seen on an oscilloscope?
4. At what frequencies do circuit parasitics, such as lead inductance, become important to measurement accuracy or circuit operation?
5. How can the effect on an electronic circuit of a measurement be minimized?

6.   What techniques are available to prevent external interference from introducing error into a measurement?

In order to answer these and other questions, the following topics will be discussed at length in this book:

1.   High frequency measurement techniques and philosophy
2.   Review of basic theory including:
       Inductance
       Magnetic coupling
       Electromagnetic interference
3.   Scope probe effects:
       Ground lead resonance
       Ground lead impedance
       Compensation
4.   Noise measurement techniques:
       Balanced coaxial probe
       Square magnetic loop
5.   Noise source location techniques
6.   General measurement philosophy
7.   Measurements on switching power supplies
8.   Measurements involving ElectroStatic Discharge, and other severe forms of interference to electronic systems.

Oscilloscope Probes

Scope probes are familiar to most engineers and electrical engineering students. This familiarity will be exploited in the following chapters as a context for introducing measurement concepts. These concepts will then be extended to more advanced measurement techniques.

The scope probe ground lead, called a pigtail in EMI circles, can result in errors on the order of 100 percent in the peak amplitude of a measurement. The inductance of this lead can cause resonant effects and inject unintended signals into the measurement. Chapter 4 will be devoted to this very important subject. The reader will be able to quantitatively measure the amount of error that results from the probe ground lead or any pigtail connection of the shield of a coaxial cable.

The problems associated with scope probe measurements do have silver linings though. It is through a discussion of these problems that a general measurement philosophy which has application to all types of measurements will be developed. This philosophy, a central point of which is to be suspicious of measurement results, will be applied to more advanced measurement techniques later in the book.

New Measurement Techniques

It is easy to be intimidated by an array of expensive test equipment. Figure 1.1 shows an effect often noticed by the author. The effect is that the confidence in a measurement is monotonically related to the cost of the test equipment. As the cost of the measurement equipment approaches \$40,000, most of us tend to believe the measurement.

MEASUREMENT CONFIDENCE FACTOR

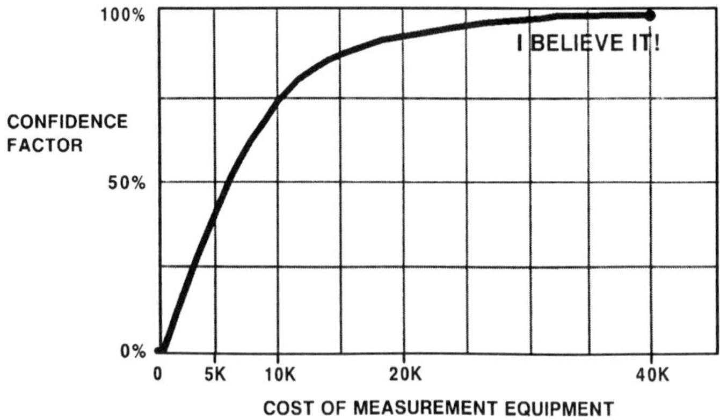

Figure 1.1 Measurement confidence factor.

A few chapters are devoted to new measurement techniques which prove the relationship of Figure 1.1 to be false. These techniques are very powerful, locating and fixing design problems in minutes where standard techniques failed after days and weeks of effort. Yet, the probes used for these techniques can be built inexpensively in a matter of minutes by the reader.

NOISE SOURCES

Since many of the measurement techniques to be discussed will be applied to noise measurements, even though they are applicable to signal measurements as well, some discussion of noise sources is warranted at this point.

General

Noise sources affecting modern electronic circuits vary widely in their characteristics. One trait that is common to all is the ability to interfere with the operation of electronic circuits and systems. Sometimes a noise source is contained within the system that is affected, intrasystem noise, and at other times a noise source in one system can affect the operation of another, intersystem noise. In addition, a third type of noise source is external to electronic systems altogether, external noise, but can nonetheless affect system operation. ESD is such an external noise source. Occasionally ESD is an internal source of noise in systems especially those with internal moving parts such as a printer.

Examples of Noise Sources

Typical sources of high frequency noise that affect electronic circuits and systems include: digital logic, switching power supplies, ESD, motors and relays, and Electromagnetic Interference (EMI) from radio transmitters. Digital logic noise is probably the most widespread form of noise in electronic systems. It can cause serious interference to broadcast and two-way radio communications. In fact, the potential for problems became so great that the Federal Communications Commission enacted rules in 1979 to limit the amount of radiated emissions that digital equipment would be allowed to emit. Also common is digital noise so severe that a circuit interferes with itself. This topic will be covered in detail throughout this book.

Switching power supplies have become very popular for their light weight and low cost. In a switching supply, the input power is typically chopped at several tens of kHz or faster and a small high frequency transformer or energy storage inductor is used to change the voltage level of the power supply. Because of the large amounts of power being switched, there is the potential for strong fields and resulting large noise voltages to be generated. Some switching supplies used in common electronic equipment are capable of generating a few volts of noise across just an inch of wire. Noise from switching power supplies, although relatively low in frequency, is a major source of noise in electronic circuits.

ESD is probably the most potent source of noise that electronic circuits may experience. An ESD event can deliver amperes of current and a voltage change of kilovolts in picoseconds of time. The shear magnitude of the resulting di/dt, dv/dt, and associated electromagnetic fields can easily disrupt circuit operation and even cause physical damage to circuit components. In addition to equipment malfunction, the presence of ESD events can affect noise measurements unrelated to ESD. ESD has assumed such an importance in the electronics industry today that there are many standards bodies as well as individual companies writing ESD testing and susceptibility standards to test electronic equipment for ESD susceptibility.

Motors, relays and switch contacts can generate sparks at points of electrical contact. This type of noise is sometimes referred to as Electrical Fast Transient (EFT). As with ESD, significant fields can be generated by EFT that can disrupt equipment operation. Often, the sparking contact is not associated with the equipment affected. This form of noise is also widespread and can be severe in its effect on electronic equipment. Some equipment which contains electromechanical components, such as printers of the type used with personal computers, can generate internal EFT and interfere with itself.

EMI generated by radio transmitters is usually continuous, seconds or longer, rather than transitory, nanoseconds or microseconds, in nature and not as strong as the interference generated by ESD or EFT. EMI is generally not as much a problem for digital electronics as it is for analog circuits. Many analog components have the ability to act like crystal radios and generate audio frequency noise from EMI. The presence of a nearby radio transmitter can also affect noise measurements made on electronic equipment.

Effects of High Frequency Noise

Noise sources, such as those described above can have many effects on electronic circuit operation such as:

1. Erosion of digital logic noise margins. The result usually has easily observable results, although not good ones.
2. Presence of noise sources within a circuit can result in EMI which affects other equipment located nearby.
3. Generation of noise at other frequencies than the frequencies contained in the original noise source, caused by the interaction of the original noise with nonlinear elements such as solid state devices.
4. Unusual malfunctions that are very difficult to track down. Many of these problems can have very low repetition rates which make location and measurement of the noise extremely difficult. Read/write errors in a magnetic storage device caused by switching power supply noise is one example of this. An error condition may only occur a few times per hour or even less frequently, but may still be a serious problem.
5. Poor reliability of equipment because of the failure of components. An example of this would be a 2 volt undershoot on a clock waveform driving an Integrated Circuit (IC). The magnitude of the undershoot violates the absolute maximum rating of the IC, possibly causing cumulative damage and eventual failure.

THE NEED FOR ACCURATE MEASUREMENTS

In today's world of ever higher speed electronics it is necessary for equipment designers to be familiar with high frequency design techniques. These are the same techniques that are necessary for making accurate high frequency measurements. Thus a familiarity with these measurement techniques will help prepare designers to work with modern high speed circuits.

Given the many sources of noise that can affect electronic equipment, there exists a need to measure, characterize, locate, and mitigate these sources as well as a need for measure and characterize signals to insure proper equipment operation in its intended environment. Accurate measurements are necessary for several reasons. Overdesign is expensive. Ground noise on a Printed Wiring Board (PWB), that has a peak amplitude of only 300 millivolts, but is measured as having a peak value of 700 millivolts, will likely result in an engineer adding cost into the circuit to fix a "non-problem."

On the other hand, if there really are 700 millivolts of ground noise on a PWB, and the measurement only shows 300 millivolts of peak noise, then the equipment may be unreliable. This type of error can lead to a costly field recall at a later date.

In addition to the above issues, equipment must often meet requirements imposed by organizations remote from the designers or users of the equipment. For digital equipment, FCC Part 15 rules on radiated emmissions must be met before the equipment can be sold. Telephone equipment may have to meet FCC Part 68 rules for connection to the telephone network as well as Part 15 rules. Some corporate customers of electronic equipment, such as computer users, require that the equipment meet either a corporate or industry standard of performance. Accurate measurements are the key to insuring compliance to any required rules and standards at the least cost.

Conventional measurement techniques employed by engineers and students can be grossly inaccurate unless time is taken to understand the measurement. Extreme errors in peak amplitude are possible, if indeed the measurement can be made at all. The measurement techniques presented in the following chapters will ensure accurate measurements of signals and noise as well as introduce the reader to a number of new techniques.

SUMMARY

In this chapter the following points were made:

- High frequency design and measurement techniques are not reliant on "magic."
- The frequency range of 20 to 500 MHz will be covered with extensions to the range of frequencies above one GHz where appropriate.
- A philosophy of measurement will be developed in this book that has application far beyond the particular techniques discussed.
- Oscilloscope probes will be used as a context for introducing measurement concepts.
- New measurement techniques will be presented that have not been previously published.
- Noise sources to be discussed include digital logic, switching power supplies, ESD, EFT, and EMI from radio transmitters.
- High frequency noise can have dramatic effects on circuit operation including physical damage to components.
- Accurate measurements of signals and noise are necessary if equipment is to function properly and meet its requirements.

LABORATORY DEMONSTRATIONS

Each chapter of the book will include questions and laboratory experiments. Easy to set up and perform experiments for the reader will be described including some questions that the experiments should answer. The experiments will model real engineering measurement situations. Experienced engineers will recognize flaws in common measurement practice while students will gain valuable experience from these experiments.

Consider the following scenarios. Each one represents a real engineering problem for which the solutions will be discussed in later chapters.

1. A hard disk of the type used in personal computers has been incorporated into a large piece of equipment containing dozens of PWBs, several switching power supplies, and several types of analog and digital interfaces to other equipmemt. A few times per hour on a continually running write/read test, a few bytes become corrupted. On an overnight run, the error rate is an order of magnitude greater than the published specifications of the hard disk. Replacing the hard disk has no effect on the problem.

    - What is a likely cause of the problem?
    - How would an engineer go about identifying the source of the problem?
    - How about a fix?
    - How can the margin, or robustness, of the fix be measured?

2. An Address Latch Enable (ALE) lead connects to a microprocessor and its memory on a PWB. Because of the nature of an observed problem, a glitch is suspected to occur on the ALE lead. Upon connecting an oscilloscope to the ALE lead the circuit starts to function properly and no evidence is seen on the oscilloscope of noise that would cause a problem.

    - By what mechanisms can the connection of measurement equipment cause a circuit to start working?
    - How can the original noise problem be measured if not with standard test equipment?
    - What information does the reaction of the circuit being tested to the connection of test equipment yield about the original problem?

3. ESD is suspected as a cause of equipment malfunction even though no ESD events have been directly observed.

    - How can the presense of ESD events near electronic equipment be detected using readily available laboratory equipment?
    - If detected, how can the source of the ESD be located?

2

Theoretical Background

GENERAL

Most of this book emphasizes practical results. However, some theoretical background is necessary for a good understanding of signal and noise measurement and noise mitigation. This chapter contains the background required to understand the following chapters. The topics to be reviewed include:

- Electric field coupling,
- Inductive voltage drop,
- Magnetic field coupling and transformer action,
- Coaxial cable operation, and
- Resistive/capacitive voltage divider.

The reader may choose to skip any topics that are familiar and continue on to Chapter 3.

ELECTRIC FIELD COUPLING

E-field, or electric field, coupling occurs because of a mutual capacitance between two conductors. The current through the mutual capacitance is related to the voltage across it by Equation 2.1 with reference to Figure 2.1.

$$I_C = C \cdot (dV_C/dt) \qquad (2.1)$$

Where I_c is the current through the capacitor and dV_c/dt is the rate of change of voltage across the capacitor.

The coupling results in a current injected into the receiving circuit and in a noise voltage, V_n, given by Equation 2.2.

$$V_n = I_c \cdot (Z_s || Z_1) \qquad (2.2)$$

Where V_n appears across the load Z_l, and Z_s is the signal source impedance in the receiving circuit. The operator $||$ means "in parallel with."

In general, V_n is much smaller than V_1 because the reactance of the coupling capacitance is very large compared to the impedance to ground of $Z_s || Z_l$, thus $V_1 \approx V_c$. With this approximation, Equations 2.1 and 2.2 can be combined to form:

$$V_n = C \cdot dV_1/dt \cdot (Z_s || Z_1). \qquad (2.3)$$

Where V_1 is the voltage to ground in the source circuit, see Figure 2.1.

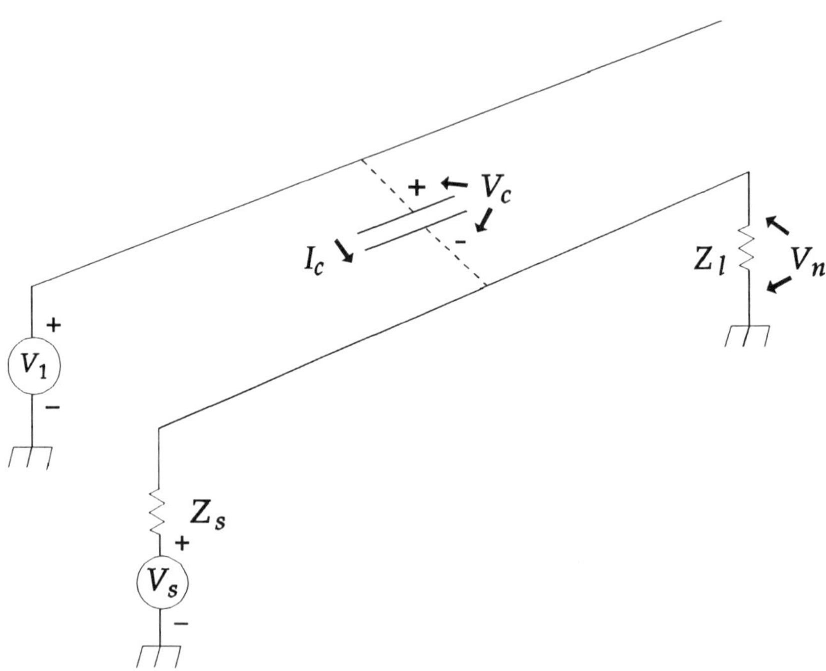

Figure 2.1 Capacitive coupling.

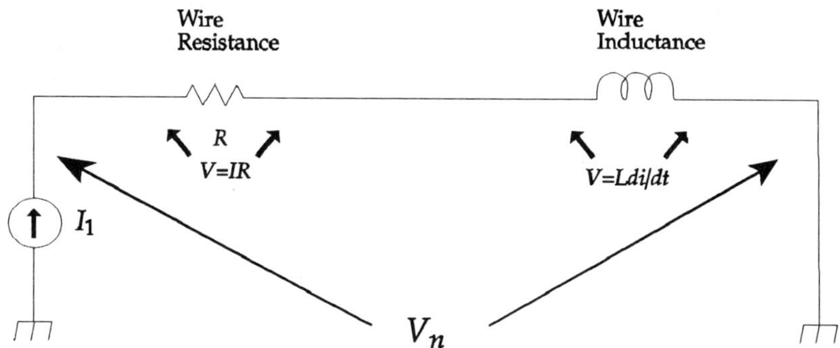

Figure 2.2 Impedance of a wire.

Since the time derivative of the sine wave source $V_1 = V \cdot e^{j\omega t}$ is $j \cdot \omega \cdot V \cdot e^{j\omega t}$ *or* $j \cdot \omega\, V_1$, equation 2.3 becomes:

$$V_n = C \cdot j \cdot \omega \cdot V_1 \cdot (Z_s||Z_l) = j \cdot 2 \cdot \pi \cdot f \cdot C \cdot (Z_s||Z_l) \cdot V_1 \quad (2.4)$$

or

$$V_n/V_1 = j \cdot 2 \cdot \pi \cdot f \cdot C \cdot (Z_s||Z_l). \quad (2.5)$$

Notice that the coupled voltage, V_n, is proportional to the value of the coupling capacitance and the frequency.

INDUCTIVE VOLTAGE DROP

Current, I_1, flowing in a conductor generates a series voltage because of the resistance and inductance of the wire. The voltage, illustrated in Figure 2.2, is given by:

$$V_n = I_1 \cdot R + L \cdot dI_1/dt. \quad (2.6)$$

If $I_1 = I \cdot e^{j\omega t}$, then $dI_1/dt = j\omega \cdot I_1$ and Equation 2.6 becomes:

$$V_n = I_1 \cdot R + j \cdot 2 \cdot \pi \cdot f \cdot L \cdot I_1. \quad (2.7)$$

Dividing Equation 2.7 by I_1 gives the impedance of the conductor:

$$Z = V_n/I_1 = R + j \cdot 2 \cdot \pi \cdot f \cdot L. \quad (2.8)$$

The definition of inductance normally is based upon a specified loop geometry.[1] For the purposes of this discussion, the inductance of a conductor is calculated by assuming the return current to be at infinity. Since magnetic field falls off quickly from a current carrying conductor, in practice the return current path need be only inches away.

At most frequencies of interest, above a few tens of kHz, the contribution to the impedance of a conductor by resistance is small compared to the contribution due to inductive reactance. Table 2.1 shows the contributions of resistance and inductance to the total impedance of several lengths of 24 gauge wire at 5 MHz as well as the DC resistance.

Table 2.1. Resistance and inductance of a wire.

Length (in)	R D.C. (ohms)	L (nH)	R at 5 MHz (ohms)	X_L at 5 MHz (ohms)
1	.004	20	.009	0.67
3	.013	60	.028	2.0
6	.025	120	.055	4.0
12	.050	240	.110	8.0

R and L are the DC resistance and the inductance of the wire respectively and X_L is the inductive reactance of the wire.

The resistance of the wire increases with frequency because of the skin effect. However, this effect is relatively weak, only doubling the resistance of the wire between DC and 5 MHz, while the much stronger effect of inductance increases the inductive reactance proportional to frequency to over 70 times the resistance at 5 MHz, 8 ohms for a 12 inch length. In fact, the inductive reactance of most common conductors usually exceeds their resistance above a few tens of kHz, by about 30 kHz in this example. Unless otherwise stated, it is assumed for the

---

1. Although normally done, it is not necessary to specify a loop geometry to define an inductance as was demonstrated by the concept of partial inductances and mutual inductances by Ruehli in the *IBM Journal of Research and Development,* Sept. 1972, pp. 470-481.

purposes of this book that the resistance of a conductor is very small compared to its inductive reactance.

MAGNETIC FIELD COUPLING (TRANSFORMER ACTION)

H-field, or magnetic field, coupling occurs because of mutual inductance between two conductors. The coupling causes a series voltage to be induced into the receiving conductor as shown in Figure 2.3. Its magnitude is given by Equation 2.9.

$$V_n = M \cdot (dI_1/dt) \tag{2.9}$$

Where V_n is the open circuit induced voltage in the receiving conductor, M is the mutual inductance between the two conductors, and dI_1/dt is the rate of change of current in the source circuit. The voltage delivered to the load, Z_l, due to I_1 is:

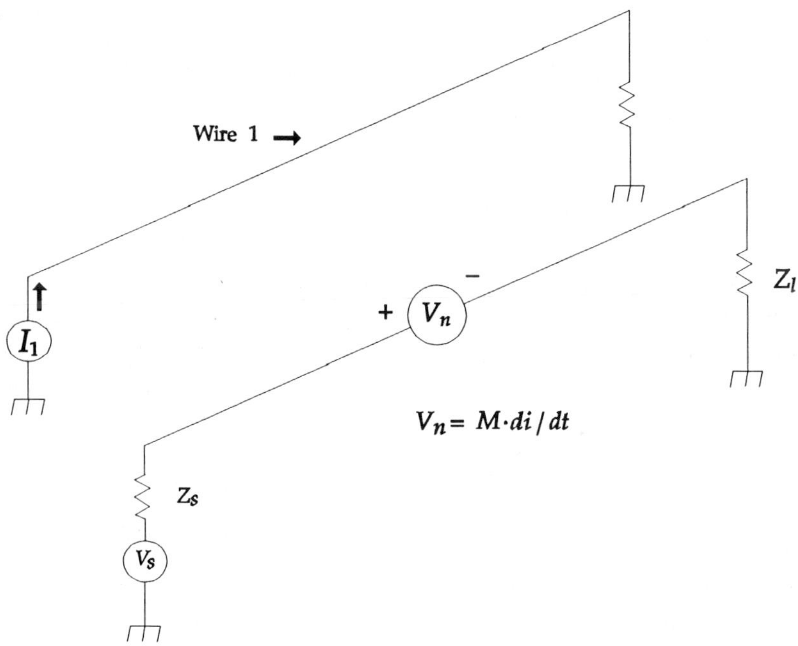

Figure 2.3 Mutual inductance.

$$V_l = V_n \cdot \frac{Z_l}{Z_l + Z_s}. \tag{2.10}$$

Where Z_l and Z_s are the load and source impedances respectively of the receiving circuit. Combining Equations 2.9 and 2.10 yields:

$$V_l = M \cdot (dI_1/dt) \cdot \frac{Z_l}{Z_l + Z_s}. \tag{2.11}$$

If I_1 has the form $[\, I \cdot e^{j\omega t}\,]$, a sinewave source, then Equation 2.11 becomes:

$$V_l = j \cdot 2 \cdot \pi \cdot f \cdot M \cdot \left[\frac{Z_l}{Z_l + Z_s}\right] \cdot I_1 \tag{2.12}$$

or

$$V_l/I_1 = j \cdot 2 \cdot \pi \cdot f \cdot M \cdot \left[\frac{Z_l}{Z_l + Z_s}\right]. \tag{2.13}$$

The quantity V_l/I_1 has units of impedance. This is sometimes called the transfer impedance between the two circuits. This concept will be especially useful in discussing magnetic pickup loops and current probes in Chapters 7 and 8.

COAXIAL CABLE OPERATION

Since shielded cables, of which coaxial cables are a subset, are widely used in high frequency design and measurement, it is important to understand how such cables work. Misuse through improper shield termination is a major source of error in high frequency measurements.

Figure 2.4 shows a hollow cylinder carrying a current uniformly distributed around its surface. Eventually this will become the shield of our hypothetical shielded cable, but for now we want to determine where the magnetic field is located for the situation of Figure 2.4. If the hollow tube is viewed end-on, as shown in Figure 2.5, each *dL* of length around the tube will be carrying the same current per unit length. To determine what magnetic field exists inside the tube, let us consider the center first.

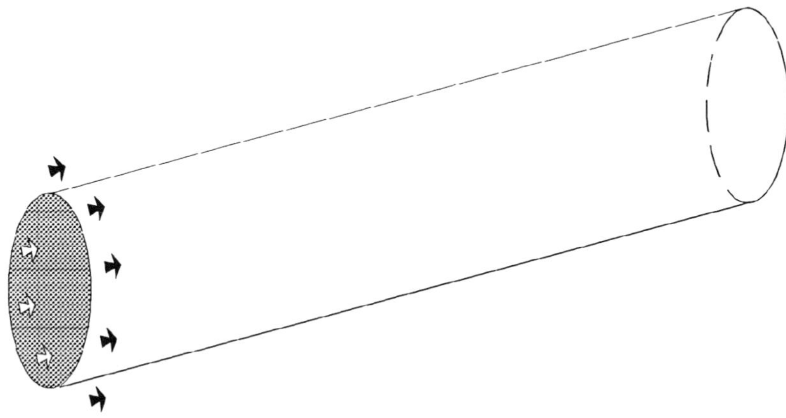

Figure 2.4 Hollow tube carrying a uniformly distributed current.

Figure 2.5a shows two equal dLs on the tube. By the Right Hand Rule, the field generated by each dL is composed of circular paths. At the center of the tube dL_1 generates M_1 and dL_2 generates M_2. Since the center is equidistant from the dL_1 and dL_2, the field strength from each is equal there. However, being on opposite sides of the tube the direction of M_1 and M_2 are opposite so they cancel and there is no net magnetic field in the center of the tube caused by the uniform current on the tube.

Now consider a point at an arbitrary location inside the tube. Consider two lines intersecting at the point, A, and crossing the tube to define dL_1 and dL_2 as shown in Figure 2.5b. As in the center, a magnetic field, M_1 is generated at point A by dL_1 and M_2 is generated by dL_2. Unlike the center case, dL_1 is not equal to dL_2.

Each dL is small enough to be similar to a long current carrying wire and as such, its magnetic field falls off as $1/r$, where r is the radius from the wire to a point. Note also that each dL and the two lines to it from point A form an isosceles triangle. The two triangles share the same angle between their equal sides. Because of this, Equation 2.14 holds.

$$dL_1/dL_2 = r_1/r_2 \qquad (2.14)$$

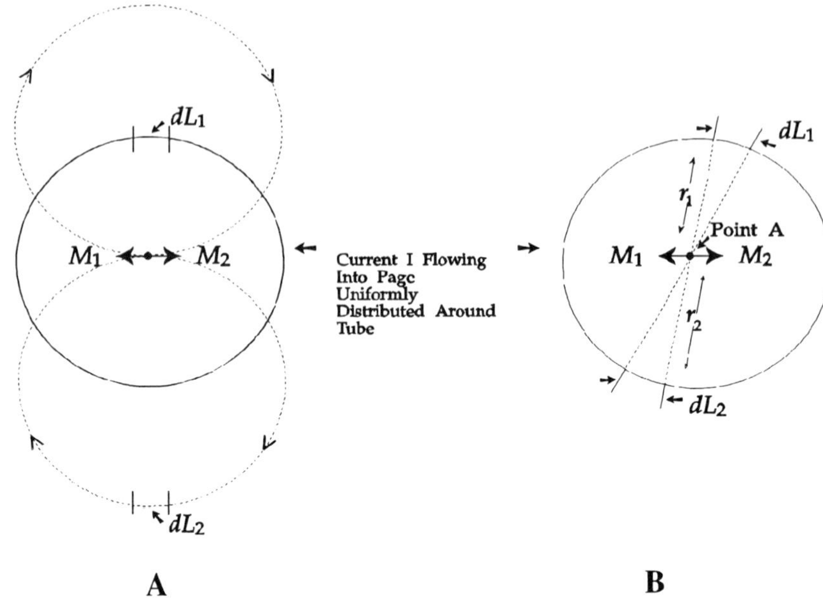

Figure 2.5 Calculating the magnetic field inside a hollow tube carrying a uniformly distributed current by graphical techniques.

Since the magnetic field from each dL falls off as $1/r$ and is also proportional to dL (because of the uniform current condition) we can write:

$$|M_i| = C \cdot dL_i/r_i \qquad (2.15)$$

where C is a constant and therefore we have the ratio:

$$|M_1/M_2| = \frac{C \cdot dL_1/r_1}{C \cdot dL_2/r_2} = \frac{C \cdot dL_1 \cdot r_2}{C \cdot dL_2 \cdot r_1}. \qquad (2.16)$$

Substituting for dL_1/dL_2 from Equation 2.14 into Equation 2.16 yields:

$$|M_1/M_2| = \frac{C \cdot r_1 \cdot r_2}{C \cdot r_2 \cdot r_1} = 1. \qquad (2.17)$$

The magnetic fields at point A, M_1 and M_2, are opposite in direction, as was the case for the center of the tube, and since their magnitudes are equal, there is no net magnetic field at point A due to dL_1 and dL_2.

If the lines passing through point A that define dL_1 and dL_2 are now rotated through 180 degrees, it can be seen that the contributions to the magnetic field at point A from all pairs of dLs are zero. Since point A can be anywhere inside the tube, there must be no net magnetic field anywhere inside the tube as long as the current is uniformly distributed around the tube. Since there must be a magnetic field somewhere, it must be outside the tube.

Another way to derive this result is to use the following relationship from Maxwell's equations:

$$\int \mathbf{H} \cdot \overline{dL} = \int \mathbf{J} \cdot d\mathbf{S} \qquad (2.18)$$

Equation 2.18 says that the net magnetic field integrated around any closed path must be equal to the amount of current enclosed by that path. In Figure 2.6 it can be seen that the net magnetic field around any path inside the tube must be zero and that there is a net magnetic

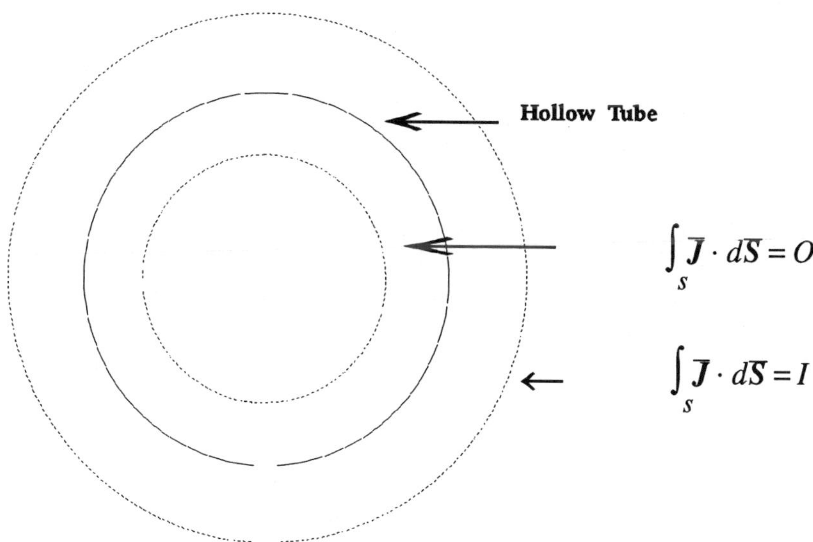

Figure 2.6 Calculating the magnetic field inside a hollow tube carrying a uniformly distributed current by Maxwell's equations.

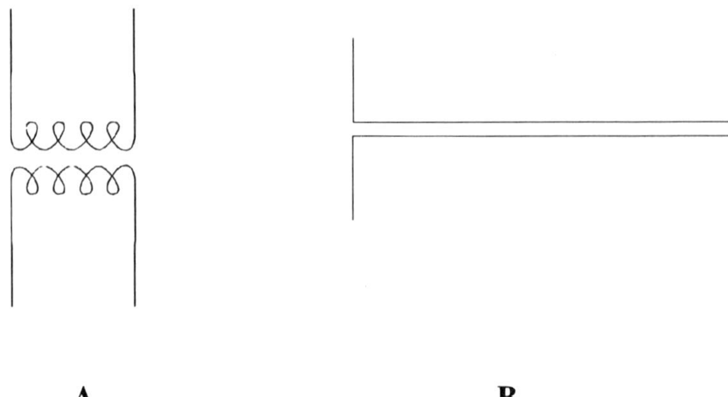

A　　　　　　　　　　　　**B**

Figure 2.7 A pair of wires as a 1:1 transformer.

field around any path outside the tube. The magnetic field lines of force are indeed circles concentric with and outside the tube.

This result has important implications for the mutual inductance between the shield and inner conductor(s) of a shielded cable. Consider the two configurations of Figure 2.7.

The transformer with a 1:1 winding ratio shown in Figure 2.7a is equivalent to the pair of wires with a parallel section shown in Figure 2.7b. In a transformer, the magnetizing inductance is the inductance of a winding with the other winding open and the leakage inductance is the inductance of a winding with the other winding shorted. For our pair of wires, the "magnetizing" inductance is just the self inductance of a piece of wire. The leakage inductance, as in the transformer, occurs because some of the magnetic flux from the primary is not enclosed by the secondary. Figure 2.8 illustrates the situation.

Source I_p drives a current around the primary circuit resulting in an induced $(L \cdot di/dt)$ voltage V_p across the "primary" or section of wire parallel to the "secondary." The flux lines generated by the primary are circles concentric to the wire. Many of them cut the loop formed by the secondary and its connections to the points where V_i, the induced voltage, is measured. These flux lines are responsible for the mutual inductance between the two wires.

Those flux lines that pass between the wires or beyond the connections to the V_i measurement do not cut the measurement loop and contribute to the leakage inductance between the wires. The flux drops off rapidly from the primary wire so the flux beyond the V_i connection can usually be neglected. Since there must be some space between the primary and secondary wires, there must be some leakage flux and thus some leakage inductance. Thus the mutual inductance, M, between the wires must be less than the self inductance of either wire, L, by itself.

A unique situation exists within a shielded cable that is illustrated in Figure 2.9. Figure 2.9 differs from Figure 2.8 in that a shielded cable replaces the primary and secondary wires used in Figure 2.8. The shield is used for a primary and the center conductor becomes the secondary. It is assumed for the purposes of this discussion that the current flowing on the shield is uniformly distributed around it.

From the discussion above, we know that the flux due to the current along the shield exists outside the shield only as shown in Figure 2.9. Therefore, if the loop closure at V_i is at least a few inches away, most all of the flux due to the shield current will cut the secondary loop and the mutual inductance, M, is nearly equal to the shield inductance, L_s. The coefficient of coupling between two circuits, k, is the fraction of flux from one circuit that cuts the second circuit. For this case, k is aproximately equal to unity. In situations like that in Figure 2.8 using discrete wires, k must be less than unity, with 0.5 or less being a typical

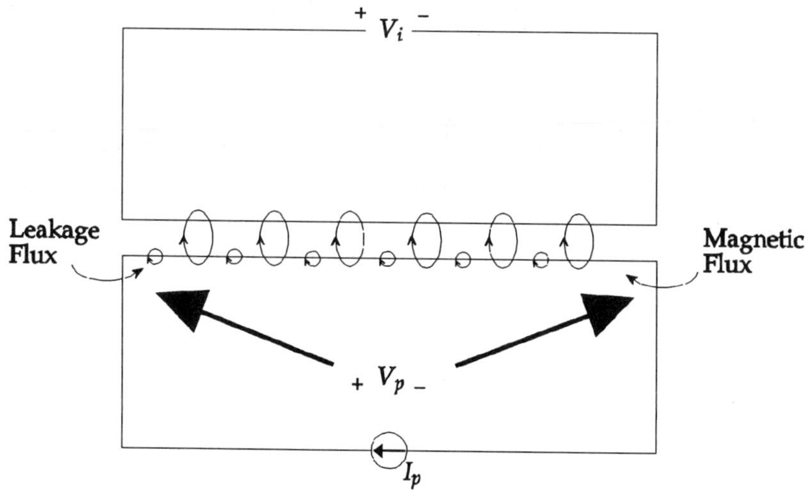

Fiigure 2.8 Magnetic coupling between two wires.

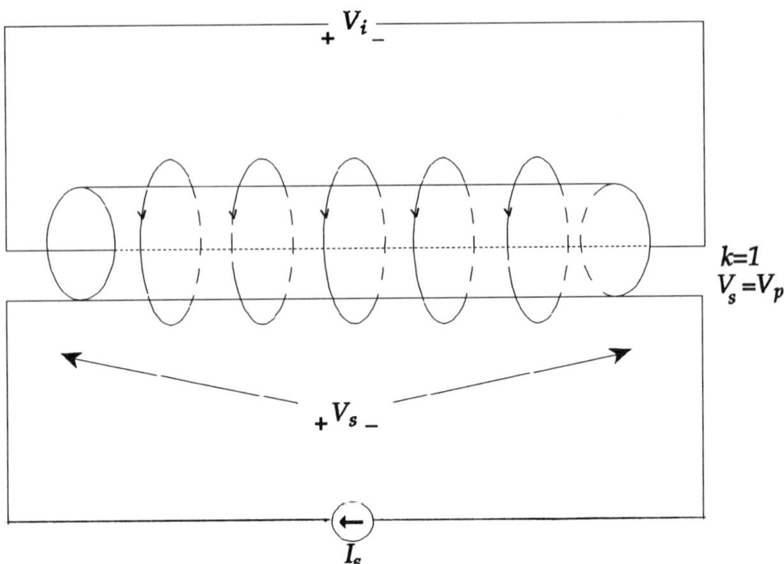

Figure 2.9   Magnetic coupling in a shielded cable.

number. Since magnetic flux falls off rapidly from a current carrying wire, as $1/r$, the measurement of V_i in Figure 2.9 need be only a few inches away for k to be very close to unity. The secondary voltage, V_i, is induced with the same polarity as V_p.

In reality, the impedance of a shield on a cable is the sum of the shield inductance, L_s, and the shield resistance, R_s. The equivalent circuit for a real shielded cable is shown in Figure 2.10. Only the voltage developed across L_s will be transferred to the center conductor. Voltage developed across the shield resistance, R_s, is not induced into the center conductor.

The fraction of the shield voltage, V_s, that is induced into the center conductor is important to the operation of the shielded cable. With reference to Figure 2.10, the induced voltage into the center conductor, V_i, can be written at a given frequency as:

$$V_i = M \cdot dI_s/dt = M \cdot j \cdot w \cdot I_s. \tag{2.19}$$

I_s can be written as:

$$I_s = \frac{V_s}{R_s + j\omega L_s}. \tag{2.20}$$

Combining Equations 2.19 and 2.20 yields:

$$V_i = \frac{M \cdot j\omega \cdot V_S}{R_S + j\omega L_S}.$$

(2.21)

Since L_S is equal to M in a shielded cable, Equation 2.21 can be rearranged as follows:

$$V_i/V_S = \frac{j\omega L_S}{R_S + j\omega L_S} = \frac{j\omega}{j\omega + R_S/L_S} = \left(\frac{1}{1 + \dfrac{R_S}{j\omega L_S}} \right)$$

(2.22)

Above the corner frequency, F_C, where $2\pi F_C L_S = R_S$, most of the shield voltage is induced into the center conductor. This corner frequency for most shielded cable types varies from a few kHz to about 100 kHz depending on the shield construction. Most measurement situations involve frequencies that are much higher than this although occasionally 60 Hz current flowing on a shield causes problems.

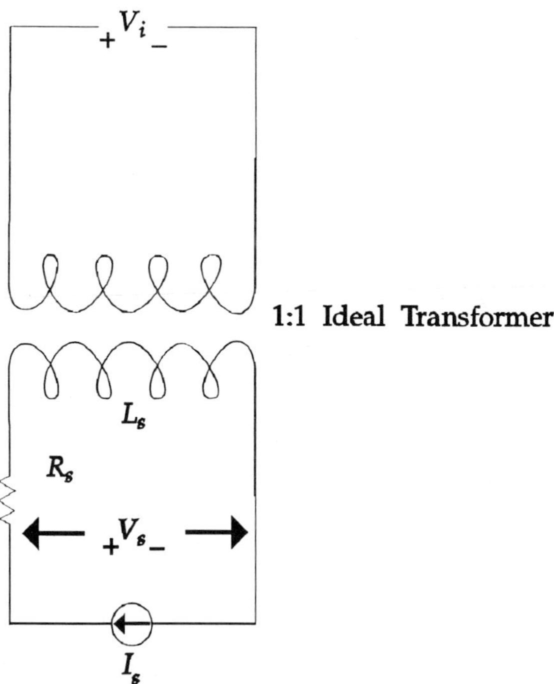

Figure 2.10 Shielded cable equivalent circuit.

The important conclusion of this discussion is this: the inductive voltage drop across the shield of a shielded cable due to current flowing in the shield will be efficiently induced onto the center conductors (there may be more than one) at most frequencies of interest.

RESISTIVE/CAPACITIVE VOLTAGE DIVIDER

A voltage divider network composed of a pair of resistor-capacitor, *RC*, networks finds its way into many engineering and measurement situations. Later in this book, application of the resistive/capacitive voltage divider will be used to explain compensation of oscilloscope probes. Frequency compensation networks, such as are used in hi fi systems, incorporate this type of network.

Consider the circuit shown in Figure 2.11. The voltage drop across R_2 and C_2, V_O, is the output of the circuit while V_{gen} is a voltage source.

The output voltage can be expressed as:

$$V_0 = \frac{R_2 \parallel C_2}{R_2 \parallel C_2 + R_1 \parallel C_1} \cdot V_{gen} \qquad (2.23)$$

Expanding Equation 2.23 results in the voltage gain, V_0/V_{gen}, becoming:

$$V_0/V_{gen} = \left(\frac{R_2}{R_2 + R_1 \left(\dfrac{1 + s \cdot C_2 \cdot R_2}{1 + s \cdot C_1 \cdot R_1} \right)} \right) \qquad (2.24)$$

where $s = j \cdot 2 \cdot \pi \cdot f$.

At low frequencies, the circuit is just a resistive voltage divider described by:

$$V_0/V_{gen} = \frac{R_2}{R_1 + R_2}. \qquad (2.25)$$

And at high frequencies the voltage gain is determined by the ratio of the capacitances and becomes:

$$V_0/V_{gen} = \frac{C_1}{C_1 + C_2}. \qquad (2.26)$$

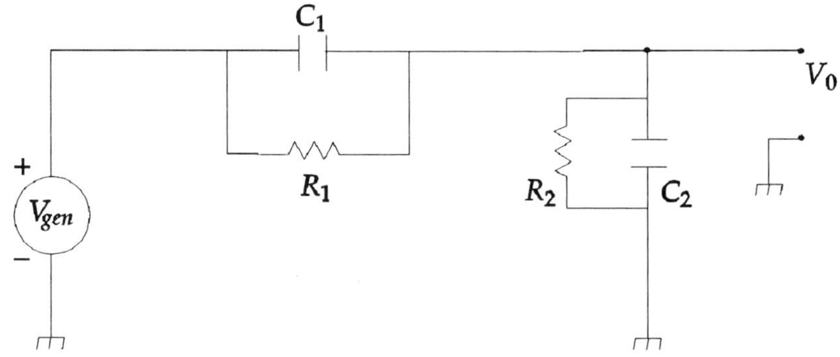

Figure 2.11 Capacitive divider.

It is evident from Equation 2.24, that for the voltage gain of the circuit to be constant over the frequency, the time constants of the circuit, $(R_1 \cdot C_1)$ and $(R_2 \cdot C_2)$, must be equal.

LABORATORY DEMONSTRATIONS

2.1.   Consider the measurement depicted in the following figure. A high frequency voltage source, V_m, is connected through a length of coaxial cable to a load, R_0, which matches the characteristic impedance of the cable. At two points along its length, the shield of the coaxial cable comes into contact with ground forming what some technical people call a "ground loop." In addition, because of other noise and signal sources, there are potential differences between each of the grounds G_1 through G_4. Thus the voltage drop between G_1 and G_4 is $V_1+V_2+V_3$.

For this example, assume that the shield terminations are such that the shield current is uniformly distributed around the circumference of the coaxial cable shield. Ignore any effects caused by the inductance and resistance of the connections to ground as well as shield resistance.

Do the two incidental connections to ground interfere with the measurement of V_m at V_{out}? Write the equation for V_{out} as a function of all the voltages in the circuit.

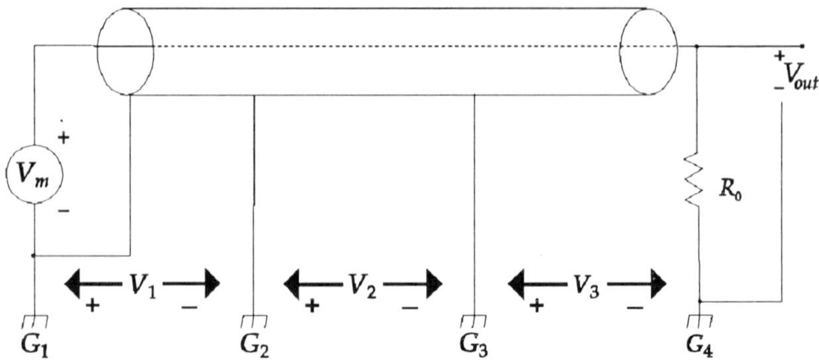

Figure for Problem 2.1 Coaxial cable with multiple shield grounds.

Answer: $V_{out} = V_m$ The extra ground connections have no effect on the measurement. Current does flow in the shield because of the ground potential differences, but these voltages are induced into the center conductor of the coaxial cable so as to cancel the ground potential differences around the measurement circuit (loop) consisting of R_0, the cable shield, V_m, and the center conductor. We can see this by adding the voltage drops around the circuit. Since the sum of the voltage drops must equal zero, the sum of all drops except V_{out} must equal V_{out}. The sum of the voltage drops is:

$$V_{out} = \underbrace{V_3 + V_2 + V_1}_{(shield\ voltage)} \quad \underbrace{+\ V_m}_{(source)} \quad \underbrace{-\ V_1 - V_2 - V_3.}_{(center\ conductor\ voltage)}$$

The center conductor voltages are induced from the shield with the same polarity, but since adding around the loop takes the shield voltages from left to right and the center conductor voltages from right to left, the center conductor voltages carry minus signs and thus cancel the shield voltages for an ideal shielded cable. Therefore: $V_{out} = V_m$ and the ground noise does not affect the measurement.

2.2. Consider the slightly modified measurement depicted in the figure below:

Voltage source, V_m, is now connected through three sections of coaxial cable to a termination. R_o is equal to the characteristic impedance of the cable. As in Problem 2.1, it is desired to use the voltage V_{out} as a measure of V_m and because of noise sources not shown, each connection to ground is at a different potential relative to the ground of V_m, G_1, as shown in the figure.

As in Problem 2.1, assume that the shield terminations are such that the shield current is uniformly distributed

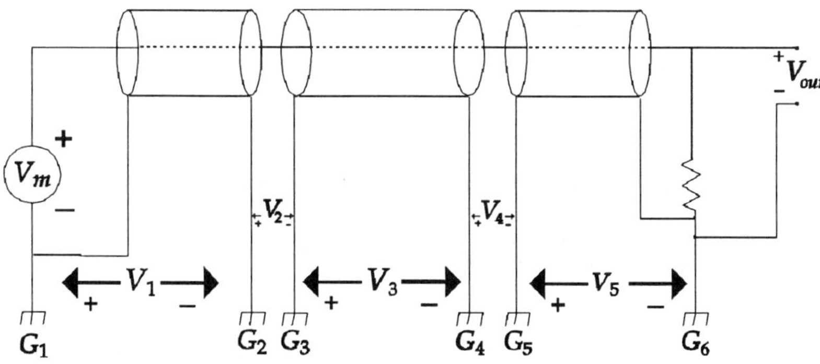

Figure for Problem 2.2 Coaxial cable with segmented shield.

around the circumference of the coaxial cable shield. Ignore any effects due to the inductance and resistance of the connections to ground as well as to shield resistance. Write the expression for V_{out}. Is there interference to the measurement for this case?

Answer: $V_{out} = V_m + V_2 + V_4$. Yes, there is interference, but V_1, V_3, and V_5 are induced into the center conductor by the shield and thus cancel out of the measurement just as in the previous example, leaving only V_2 and V_4 contributing to the interference. In fact, sources like nearby ESD can cause the peak values of V_2 and V_4 to far exceed the peak value of V_m. For instance, if grounds G_2/G_3 and G_4/G_5 are only a few inches apart, an ESD event several feet away could result in V_2 and V_4 having peak values of several volts or more.

2.3.   The figure below depicts a measurement using a single coaxial cable. The source, V_m, is delivered into R_o, a match for the characteristic impedance of the cable, resulting in an output voltage, V_{out}.

Assume the ground lead between V_m and the cable shield has an inductance of 80 nanohenries, about 4 inches of wire. Due to an ESD event across the room, an impulsive potential, V_1, causes a current, I_1 to flow on the ground lead and onto the shield of the cable as shown. I_1 reaches a value of 2 amperes peak in 10 nanoseconds. The ground lead inductance in the connection to R_o is considered negligible and the coaxial cable is to be considered ideal with uniform shield current.

- What is the potential error in the measurement of V_m at V_{out}?
- How important is shield resistance for this case?

Answer: $V_{out} = V_m + V_g$ where V_g is the voltage across the cable ground lead at V_m. V_g has a magnitude of:

$$V_m = (80 \text{ nH} \cdot 2 \text{ A}/10 \text{ ns}) = 16 \text{ volts!}$$

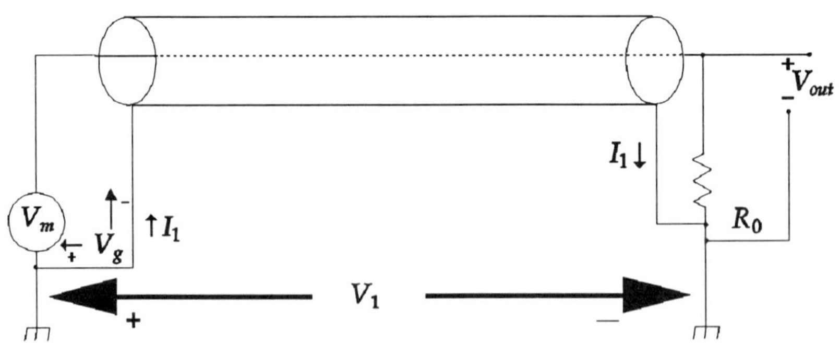

Figure for Problem 2.3 Coaxial cable with pigtail connection of shield.

This is not an unreasonable value in real life measurements. ESD events can cause very high *di/dt values in nearby equipment. Since the frequencies involved in this example are much higher than 100 kHz, normal values of shield resistance will be small compared to the shield inductive reactance and therefore will not affect the measurement substantially.*

3

Practical Background/Probes

INTRODUCTION

Before discussing measurement techniques, it is appropriate to cover some practical background information as well as some of the material that will be needed, in future chapters, for laboratory experiments. The topics covered will include scope probes, current probes, and a useful signal generator to be used in the laboratory experiments in the chapters to follow. Probes, their operation, types available, and limitations will be discussed in detail. The reader may proceed to Chapter 4 and refer back to this chapter for reference if needed.

OSCILLOSCOPE PROBES

Oscilloscope measurements will be used as a platform to discuss measurement concepts in later chapters, so an understanding of scope probes and their operation is crucial to the usefulness of later chapters. Two major types of probes available are active probes (containing electronic circuitry with active components, usually FETs) and passive probes. Within the passive probe category, there are high impedance and low impedance probes. High impedance passive probes are the most popular type of oscilloscope probes. These include the ubiquitous 10X probes that most students and engineers find all too familiar. It is this familiarity that is a major source of error in measurements, for little thought is usually given to parasitic effects of the probe operation and use. These parasitic effects can, under some circumstances, lead to significant errors.

High Impedance Passive Oscilloscope Probes

A typical 10X high impedance probe uses an RC network at the probe tip to form a 10:1 voltage divider with the scope input impedance and the cable capacitance. Typically the RC probe tip network might be approximately a 9 megohm resistor in parallel with a 12 pF capacitor. A typical scope load would be one megohm in parallel with 20 pF.

The probe cable is of a special design. The characteristic impedance, Z_0, is made as high as possible with a typical value being 170 ohms. Even so, the cable cannot be matched at the scope end if it is to be a high impedance probe. Reflections and spurious response caused by the termination of the probe cable in one megohm in parallel with about 20 pF at the scope is controlled by making the center conductor lossy, typically about 39 ohms per foot. Sometimes a network that contains time constant adjustments and inductive peaking is used at the scope end of the probe to help insure a flat response. The cable capacitance, typically 8 pF per foot, appears in parallel with the scope input impedance, adding to its capacitance up to a few tens of MHz. Above that frequency the scope cable begins acting more like a transmission line.

Since the probe cable is not terminated in its characteristic impedance at the scope, there will be some reflections at higher frequencies and the resultant ringing and spurious response despite efforts to minimize them. The spurious response of a probe is part of the probe specification. However, this specification is only valid when the probe bandwidth is matched to the scope bandwidth. If a scope of 350 MHz bandwidth scope is used with a 100 MHz probe, the probe's spurious responses will be displayed with greater amplitude, resulting in a potentially less accurate measurement. This result is counter to our intuition which suggests that improving the bandwidth of one component of a system will improve the system response.

There is a catch-22. Probe bandwidth is rarely printed on the probe itself and the instruction manual for the probe is generally lost soon after the probe is purchased. In addition, many laboratory oscilloscopes do not have their vertical bandwidth printed on the instrument either. The probability of matching the scope bandwidth to the probe bandwidth approaches zero in a typical engineering laboratory equipped with several probes and oscilloscopes, none of which is labeled for bandwidth.

Input Impedance

Another problem with high impedance passive probes is that the input capacitance of 10 pF or so can become a fairly low impedance at frequencies above 50 MHz. At 50 MHz, 10 pF has a capacitive reactance of only about 300 ohms. As will be shown in the next chapter, this capacitive reactance is a major source of measurement error.

At very low frequencies, lower than a few kHz, the input impedance of a high impedance passive probe approaches the DC value, typically 10 megohms. In this frequency region, the probe becomes sensitive to electric field coupling to conductors connected to the probe tip. An example of this is the common 60 Hz signal seen when the probe tip is touched by an ungrounded person. Sensitivity to spurious signals such as this can affect accuracy of measurements or make some measurement results difficult to obtain. Implications of this effect will be discussed in some detail in later chapters in several measurement contexts.

High Impedance Passive Probe Compensation

As mentioned earlier, the probe tip of a 10X passive high impedance probe contains an RC network that, in combination with the scope input impedance and the cable capacitance, forms a 10:1 voltage divider. Probe compensation, adjusting the probe to the scope input capacitance, must be done with high impedance passive probes or significant errors can occur. Some probe construction techniques and circuitry can produce errors of 50 percent. Depending on the probe type, the compensation adjustment may be located in the probe tip body or in a small box at the scope end of the probe cable. A few 10X high impedance probe designs even include adjustments at both ends of the probe.

Chapter 5 will deal with the compensation of 10X passive high impedance probes in greater detail. For the purposes of this discussion, it is sufficient to note that significant amplitude and wave shape errors can occur in a 10X passive high impedance probe when it is improperly compensated. Thus it is imperative that the compensation of probes of this type be checked before each use. Some other probe types discussed in this chapter have the advantage that they do not need any compensation adjustment at all and thus eliminate a probable source of error.

Low Impedance Passive Oscilloscope Probes

A typical low impedance passive probe is a simple and effective measurement device. Such a probe is usually constructed with a length of 50 ohm coaxial cable and an input resistor as shown in Figure 3.1

The cable must be terminated in the characteristic impedance of the cable at the scope end, usually 50 ohms. Since the cable is

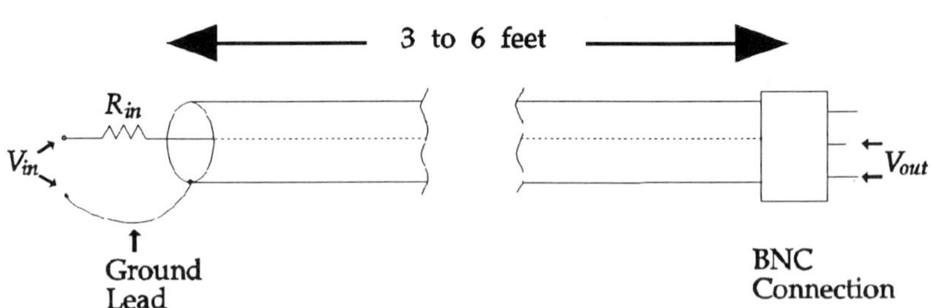

Figure 3.1 Low impedance passive probe.

terminated in its characteristic impedance, Z_o, the impedance looking into the input end of the cable is Z_o over a wide range of frequencies.

The value of the input resistor, R_{in}, sets the probe gain and its input impedance. The probe gain is just the divider ratio of R_{in} and Z_o as given by:

$$\frac{V_{out}}{V_{in}} = \frac{R_{in}}{R_{in} + Z_o}. \tag{3.1}$$

The probe input impedance is simply $(Z_o + R_{in})$. So if 50 ohm cable is used and a 10:1 probe is desired, R_{in} must be 450 ohms and the probe input impedance will be 500 ohms. A 10:1 probe with 500 ohms input impedance is a popular combination and there are available many commercial versions of such a probe.

There is nothing magic about a 10:1 attenuation ratio. If the oscilloscope used has adequate sensitivity, higher ratios may be used to achieve a higher input impedance. For instance, using an R_{in} of 950 ohms with a 50 ohm cable results in a 20:1 probe with a 1000 ohm input impedance. Ratios as high as 50:1 or 100:1 are practical in many measurement situations with some limitations.

Limitations of the Low Impedance Passive Probe

This type of probe has two main limitations, assuming the cable itself to be of high quality. First, R_{in} has a parasitic capacitance, C_{in}, in parallel with its resistance. This capacitance is partly due to the body of the resistor itself and partly to the capacitance between the leads of the resistor. At a high enough frequency, the magnitude of the capacitive reactance of C_{in} equals R_{in}, lowering its impedance substantially. That frequency is given by:

$$F_c = \frac{1}{2\pi \cdot R_{in} \cdot C_{in}}. \tag{3.2}$$

Above this frequency, the gain of the probe increases with frequency over a range of frequencies and the input impedance similarly decreases with frequency, eventually to a value of Z_o.

How the gain varies with frequency can be seen by realizing that a low impedance probe with its input resistor paralleled by a capacitor is a special case of the dual RC voltage divider discussed in

Chapter 2. The voltage divider ratio for Z_0 equal to R_2, R_{in} equal to R_1, and C_{in} equal to C_1 as given by:

$$V_o/V_{gen} = \left[\frac{R_2}{R_2 + R_1\left[\dfrac{1 + s \cdot C_2 \cdot R_2}{1 + s \cdot C_1 \cdot R_1}\right]}\right] \qquad (2.24)$$

where $s = j \cdot 2 \cdot \pi \cdot f$.

For this case C_2 equals zero and using the component names of Figure 3.1, Equation 2.24 reduces to:

$$V_{out}/V_{in} = \left[\frac{Z_0}{Z_0 + R_{in}\left[\dfrac{1}{1 + s \cdot C_{in} \cdot R_{in}}\right]}\right] \qquad (3.3)$$

Below the corner frequency $1/(2\pi \cdot C_{in} \cdot R_{in})$ the probe behaves just like a voltage divider composed of Z_0 and R_{in}. Above the corner frequency, the gain approaches unity.

Another limitation arises from the oscilloscope input impedance. Many scopes have an input resistance of 1 megohm in parallel with some capacitance, typically 20 pF or so, while others can additionally select a 50 ohm input resistance with a very low parallel capacitance, usually much smaller than 20 pF. Since a low impedance probe using a 50 ohm cable, as is typical, must be terminated in 50 ohms, an external 50 ohm load must be used with a scope that has only a 1 megohm input. For this case the input impedance will be 50 ohms resistive in parallel with the scope input capacitance.

At high enough frequencies, the scope input capacitance has two effects. First, it causes a mismatch of the cable characteristic impedance and resultant reflections cause waveform distortion. A reflected wave is generated which travels back on the cable toward the circuit under test. This reflection will have inverted phase from the original signal since the scope input impedance is lowered by the input capacitance. The probe tip resistor, 450 ohms for a 10:1 probe, is also not a match for the cable either so the signal reflected from the scope input reflects again from the probe tip. The probe tip impedance is higher than the cable Z_0 so there is no phase inversion of the reflected signal at the probe tip. When the signal arrives back at the scope it has undergone one inversion as illustrated in Figure 3.2. The effect

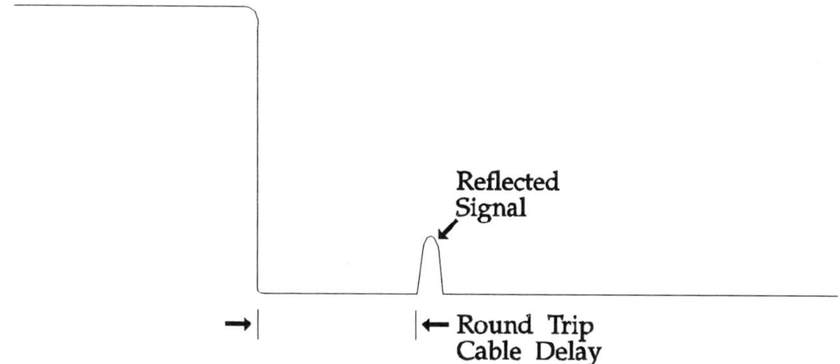

Figure 3.2 Effect of cable reflections on a fast falling edge.

is to put a spike in the displayed waveform in the direction opposite to a sharp edge and delayed in time from the sharp edge by twice the propagation time of the probe cable. For a typical 6 foot cable the delay time would be (1.8 *ns/ft · 6ft*), or about 11 ns.

For noise measurements on an electronic circuit, the peak-to-peak noise voltage is usually the important parameter. The spikes resulting from the scope input capacitance are not likely to affect this measurement significantly and can be ignored in many instances.

One possible solution is to terminate the probe end of the cable in 50 ohms so that the reflection is absorbed there. The impedance looking into the cable is now 25 ohms (Z_o in parallel with 50 ohms) so the probe has become a 20:1 probe with a 500 ohm input resistance if the probe tip resistance is 475 ohms. Figure 3.3 illustrates this case.

One way to build a 50 ohm termination impedance at the probe end of the 50 ohm cable is to take a few higher valued resistors that in parallel have 50 ohms of resistance and place them symmetrically around the coaxial cable, to reduce the inductance of the 50 ohm load, connected between the center conductor and the shield. Figure 3.4 shows such a construction.

Even if the probe tip end of the cable is matched as in Figure 3.4, another effect of scope input capacitance can be important to measurement accuracy. This effect is a loss of energy at high frequencies that occurs because the cable termination, usually 50 ohms, is paralleled by the scope input capacitance. This energy loss is a

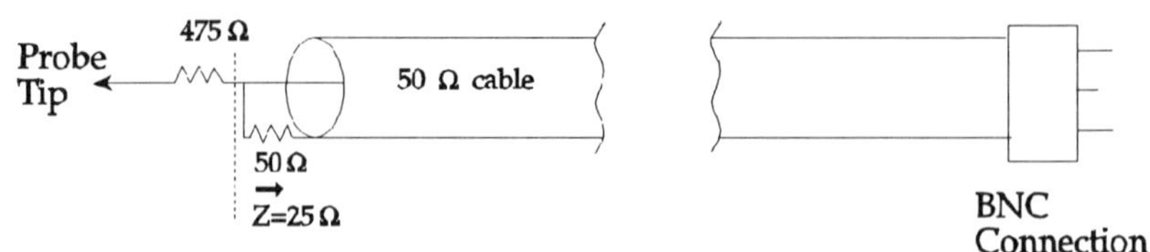

Figure 3.3 Low impedance probe with matching impedance at probe tip end.

manifestation of the reflection that occurs because of the mismatched probe cable.

The corner frequency, F_C, above which the scope input capacitance creates significant error is given by:

$$F_C = \frac{1}{2\pi \cdot R_{term} \cdot C_{scope}} \qquad (3.4)$$

where C_{scope} = scope input capacitance, and
R_{term} = cable termination resistance (usually Z_o).

Normally, above this frequency the probe response falls off at 6 dB/octave. In some cases, it is possible for this extra loss to offset the extra gain that occurs because of the capacitance of the probe tip resistance discussed above, resulting in an extended bandwidth beyond what would be expected. Probe designs usually minimize the probe tip resistor capacitance and assume connection to a high quality 50 ohm load.

Input Impedance

The input impedance of low impedance passive probes is dominated by the probe tip resistor. A 10:1 probe using a 450 ohm probe tip resistor will have close to a 500 ohm resistive input impedance up to the corner frequency, given by Equation 3.2, of the probe tip resistor, R_{in}, and its capacitance, C_{in}. For typical resistor designs, this corner frequency ranges between 300 to 500 MHz for a 450 ohm resistor. Some commercial probes have achieved flat response well beyond 500 MHz by careful design of the probe tip resistor.

The low impedance passive probe is well suited for measuring power to ground and ground to ground noise. An input impedance of 500 ohms should be much higher than the impedances encountered in these circuits. If this condition is not met, the circuit would not likely function in the first place.

This type of probe should not be used for measurements in high impedance circuits. Most high frequency circuits tend to have low impedances on the order of a few hundred ohms or less. An example of a high impedance circuit that should not be probed with a low impedance passive probe is a high Q (low loss) LC tank circuit near resonance. The input impedance of the probe can substantially lower the Q of the tank circuit and therefore the response of the tank circuit to signals.

Unlike the passive high impedance probe, the input impedance of the low impedance passive probe is resistive over a wide frequency range, and can have a significantly greater magnitude as well. In fact, the 500 ohm input impedance of a 10X passive low impedance probe is greater than the input impedance of a passive high impedance probe, that has 10 pF of input capacitance, above 50 MHz. The resistive input impedance all but eliminates resonant effects of probe input capacitance combining with the external measurement circuit. This topic will be discussed in detail in the next chapter.

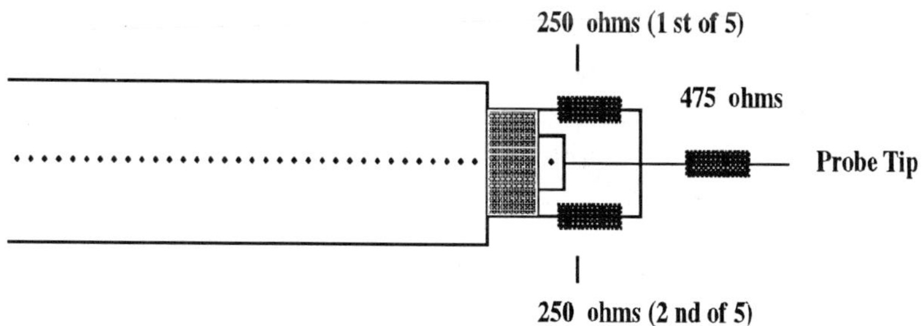

Figure 3.4 Low impedance probe with matching impedance at probe tip.

At very low frequencies, below a few kHz, low impedance probes are less sensitive to high impedance electric field coupling to the measurement circuit than are high impedance probes, active or passive. This can be a real advantage for some measurement situations. Examples of this type of measurement will be discussed in detail in later chapters.

Construction and Calibration

Unlike other probe types, construction of a high bandwidth low impedance passive probe is relatively easy. Using readily available 50 ohm cable and resistors, probe bandwidths of 500 MHz are easily achieved. When constructing measurement apparatus such as this probe, the reader should always verify that the gain and frequency response are correct. For the low impedance passive probe, a spectrum analyzer with a tracking generator output is the preferred method. Connect the probe output to the spectrum analyzer input and the probe tip and ground lead to the tracking generator output, keeping the ground lead as short as possible. A frequency range of zero to one GHz should be used. Attempting to build a probe with response flat to higher than one GHz is difficult and requires careful design and consideration of parasitic effects, such as the capacitance of the probe tip resistor.

High Impedance Active Oscilloscope Probes

There are commercially made high impedance active scope probes available in both balanced and unbalanced designs. Balanced probes have the advantage of common mode rejection. Sometimes known as FET probes, they can have high impedance inputs with only 1 or 2 pF of input capacitance that results in minimum circuit loading.

Active probes usually include an FET amplifier and power supply. Some popular designs locate the amplifier in the probe head close to the tip to minimize input capacitance. Care must be taken that the power supply is turned on before connecting the probe to the circuit. Connection while unpowered can damage some probes.

Expense can be high for this type of probe. A popular balanced FET probe sells for over $2000. In addition, these probes tend to be fragile and easily broken. One model can cost $500 to repair and the repair takes several weeks. It has been said that in order to use this type of probe one needs to purchase two. One to use while the other is in for repair. A newer design is available with modular construction and built-in input protection that reduces the possibility of damage to the probe. The modular construction allows the user to replace damaged parts to save the time and expense associated with returning the probe for service. Input protection comes at the price of increased input capacitance, about 7 pF, almost as much as a standard 10X high impedance passive probe.

Bandwidth limitations are also a concern for some high impedance active probes, especially balanced ones. A popular balanced active FET probe has a bandwidth of only 100 MHz, a bandwidth easily topped by a homemade low impedance passive probe. If an unbalanced probe is acceptable, high bandwidths are achievable. One commercially available unbalanced high impedance active probe has a bandwidth of 3 GHz and an input capacitance of only 0.7 pF.

In conclusion, the active high impedance probe, particularly the balanced variety, results in minimal circuit loading in most cases and the active circuitry allows low input capacitance. For measurements requiring very low input capacitance and low DC loading for the 500 MHz to 1 GHz frequency range, unbalanced active probes are about the only types usable for this purpose. The disadvantages of high cost, fragile construction on some models with the resultant low availability, and limited bandwidth make the balanced high impedance active probe impractical for many laboratory environments.

CURRENT PROBES

Current probes generate an output voltage that is proportional to the current flowing through an opening in the probe. Some current probes have a section than can be opened so that the probe can be clamped around a conductor without having to cut the conductor. Figure 3.5 shows a simplified clamp-on current probe used for

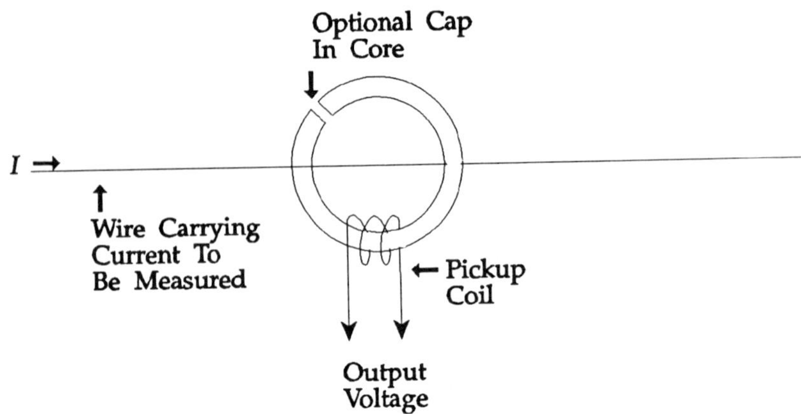

Figure 3.5 Simplified current probe construction.

measuring high frequency currents. For the purpose of this discussion, only high frequency current probes operating in the radio frequency range (1 MHz to tens of MHz) and higher will be considered. Specifically, 60 Hz clamp on type of current measuring instruments will not be discussed.

There are two mechanisms used by current probes to generate an output voltage in response to a current flowing through the probe. The most common current probes use inductive coupling to a coil in the probe to generate an output voltage. This design is basically a transformer whose primary is the wire passing through the current probe and the secondary is a pickup coil inside the probe to sense the magnetic field of the primary. A typical design uses a ferrite core around the conductor to be measured, and the pickup coil is wound around the ferrite core. This is the type shown in Figure 3.5.

The second type of current probe uses the Hall Effect to produce its output voltage in response to the magnetic field of the circuit to be measured. Outwardly, the appearance of a Hall effect probe can be similar to some inductively coupled transformer probes, but the principle of operation of a Hall effect current probe is quite different. A semiconductor is exposed to the magnetic field of the conductor to be measured and generates an output voltage in response to it. An amplifier in the probe is then used to increase this voltage to a usable value and to buffer it. A characteristic of this type of current probe is that it has response down to DC. A popular commercial probe of this type has response from DC to 50 MHz. Response to DC can be both useful and a source of error in measurements as well. This will be discussed in detail in Chapter 8.

Both the inductively coupled and Hall effect probes effectively isolate the measurement equipment from the circuit under test since no direct connection to the circuit is made. This isolation is a major advantage over scope probes of the type discussed earlier in this chapter. Connection of a probe with a ground wire to a circuit under test can result in substantial ground currents flowing on the probe that can affect circuit operation. This is called "ground loading" and use of current probes is an excellent way to eliminate it. The concept of ground loading will be discussed in detail in Chapter 4.

Current Probe Specifications

A well designed current probe should have minimal impact on circuit operation, specifically the series impedance injected into the circuit by the probe and the capacitance between the circuit under test and the current probe should be small. Injected series impedance is usually specified to be small number, 1 ohm for example. Capacitance between the probe and circuit under test is typically low, on the order of a few pF or less. These specifications, if available, can allow the user to determine what, if any, effect on circuit operation will result from using the probe.

All current probes have a specification called "transfer impedance." If the current probe output voltage is divided by the sensed current, a quantity with units of volts per ampere or ohms results. This number is called the transfer impedance of the probe. Usually current probes are specified to have a transfer impedance when their output is terminated in a specified load, usually 50 ohms. Sometimes a current probe transfer impedance is given in dB ohms, that is $20 \cdot log(Z_t)$ or dB relative to 1 ohm. Occasionally, a probe specification gives the transfer admittance, the inverse of transfer impedance.

EXAMPLE

3.1   If a probe has a 5 ohm transfer impedance then a current of 10 milliamperes flowing through the probe will generate an output voltage of 50 millivolts.

Termination of the current probe in its specified load impedance is very important. Not only does it affect the accuracy of the probe, but in the case of inductively coupled probes, termination impedance affects the impedance injected in series with the circuit under test. This is because inductively coupled current probes are really transformers and the impedance looking into the primary can be a strong function of the impedance across the secondary. Some inductively coupled current probes use internal load impedance to help limit the effect of an open circuited output. Operation of inductively coupled current probes will be discussed at length in Chapter 8.

Current Probe Electric Field Response

Although current probes are intended to sense current flowing through them, they have, to some extent, spurious electric field response. This response to electric fields is rarely included in specifications, but can have a significant impact on the accuracy of a measurement. The error is greatest when measuring high impedance circuits, those that have high signal voltages and relatively low signal currents. Measurements near ESD events tend to have this characteristic. For Hall effect probes, the response to DC can cause special problems when there are electrostatic fields present.

SUMMARY

In this chapter, several probes for measuring voltages and currents in electronic circuits were discussed. Included were three types of oscilloscope probes. These include: high impedance passive, high impedance active, and low impedance passive probes.

High impedance passive probes are the most popular and are characterized by a capacitive input impedance. These probes must be compensated to the input impedance of the scope used. Scope and probe bandwidth must be matched for the probe to meet its specifications. The high impedance passive probe can be sensitive to high impedance electric field coupling to measurements below a few kHz.

High impedance active probes are useful where circuit loading must be minimized. They are characterized by both high input resistance, above 1 megohm, and low input capacitance, as low as a couple of pF or less for some current designs. However, these probes can be fragile and expensive. Balanced versions also have bandwidth that is usually lower than that available in passive probe designs. The high impedance active probe is also sensitive to high impedance electric field coupling to measurements. Because of the low input capacitance, this effect may extend to frequencies higher, typically an order of magnitude higher, than for passive high impedance probes.

Low impedance passive probes are useful in many measurement situations and are characterized by a resistive input impedance over a wide range of frequencies. They are easy to construct for accurate response to 500 MHz, although some commercial versions are accurate to much higher frequencies. Above 50 MHz, the input impedance of this type of probe usually exceeds that of the high impedance passive probes. Low impedance passive probes are not very sensitive to electric field coupling into measurement circuits at very low frequencies.

LABORATORY DEMONSTRATIONS

Simple Signal Generator

Laboratory demonstrations for this and many other chapters in this book require the use of a square wave signal source ranging from 5 to 50 MHz to demonstrate the principles discussed. Instead of using an expensive laboratory signal generator, a simple signal generator is described below that uses an 74HC240 octal inverting buffer. Besides keeping the cost of the demonstrations low, this simple signal generator drives home the point that many of the high frequency effects discussed in this book can be seen in standard logic circuits. Most of the demonstrations described in the rest of the book use this signal generator.

Figure 3.6 shows the circuit of the signal generator. Only one half of the chip is enabled and the four unused inputs are grounded. Two buffers, B_1 and B_2, are used to implement the oscillator circuit and two more, B_3 and B_4, are used to buffer the oscillator output for two outputs. One of the two outputs, from B_4, has a 50 ohm resistor connected in series to protect the buffer output stage for the demonstrations that involve short circuiting the output. The second output, from B_3, is direct with no series resistance. BNC sockets are used for the output.

Power is derived from 4 AA alkaline batteries. Diode D_1 is used to drop about 0.75 volt from the 6 volt battery. Resistor R_6 and light emitting diode, D_2, show that power is turned on and the approximate condition of the batteries. Capacitor C_6 provides some bulk filtering of the supply voltage. Its value is not critical and 25 microfarads was chosen as a convenient value. C_7 and C_8 are high frequency bypass capacitors of the type used for DRAM, dynamic random access memory. They should be connected between V_{cc} and ground with leads which are as short as possible. Capacitor C_7 is located near C_6

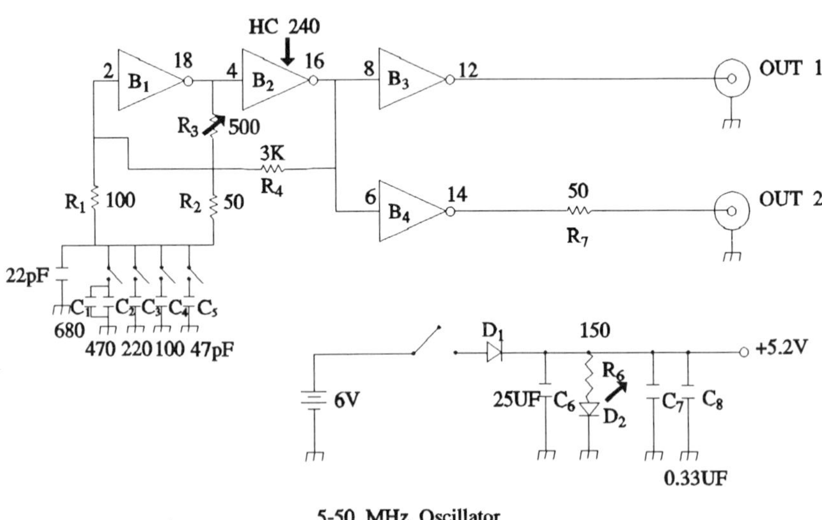

Figure 3.6 5-50 MHz Oscillator.

and C_8 is the bypass capacitor for the HC240 buffer. More detail will be given below on physical construction techniques.

Buffer B_1 is used as the oscillator. When the output of B_1 goes high, capacitors C_1 through C_5, which are switched in, charge up through R_3 and R_2. When the capacitors have charged past the input threshold of B_1, the output of B_1 goes low. Now the switched in capacitors of C_1 through C_5 discharge through R_3 and R_2 until the input of B_1 falls below its threshold. At that point the output of B_1 goes high and the cycle repeats. The time constant of $R_3 + R_2$ and the total capacitance switched in from capacitors C_1, C_2, C_3, C_4, and C_5 mainly determines the frequency of oscillation. The output of buffer B_2 is connected through R_4 to the input of B_1 to provide hysteresis at the input of B_1. This causes the capacitors C_1 through C_5 to charge to higher voltages and discharge to lower voltages during each cycle. The effect is to lower the frequency of oscillation for a given total capacitance of C_1 through C_5.

Since relatively fast rise times and high repetition rates exist in the circuit, physical layout is important. The individual buffers in the HC240 chip can produce rise times on the order of 3 ns and currents of 50 ma. Inductive voltage drops, $L \cdot di/dt$, on the order of 0.5 volt per inch of path length can be produced as a result.

Lead lengths should therefore be minimized. Particularly critical are the leads of capacitors C_1 to C_5. Inductive drop in the connections to these capacitors can cause the circuit to malfunction. The total lead length on capacitors C_2 to C_5, including the path through the dip switches, should not exceed one inch. C_1 is used to minimize circuit sensitivity to lead inductance in C_2 to C_5 by bypassing the inductive spikes across the connections to C_2 through C_5, including the dip switches. The leads of C_1 should be $\frac{1}{4}$ inch or less.

This oscillator provides more than adequate performance for the laboratory experiments in this book. It should be able to oscillate below 5 MHz on the low end and to over 50 MHz on the high end. With output rise times on the order of 3 ns and output currents of several tens of milliamperes, the circuit can generate adequate noise for the experiments in this book. Battery drain is low enough that a set of four AA batteries will last many hours of use.

The output wave shape will not be perfectly smooth and symmetrical. It will contain small bumps along the top portion of the square wave and its rise and fall times will not be exactly equal. These imperfections are caused by parasitics of the physical design and by the IC design. They will be useful for a number of experiments in future chapters by allowing inverted waveforms to be recognized.

LABORATORY DEMONSTRATIONS

Experiments

Equipment needed:

- 100 MHz oscilloscope,
- 10X passive high impedance probe,
- 10X passive low impedance probe (500 ohms),
- 1 kHz square wave generator with 50 ohm output impedance (many scopes have this for probe compensation), and
- 74HC240 based 5-50 MHz signal generator described above.

Experiment 3.1: Probe Ground Lead Effects

1. With the 10X high impedance probe properly compensated and connected to the 5-50 MHz oscillator, compare waveforms for several probe ground lead lengths up to one foot at frequencies from 5 to 50 MHz.
2. Use a 500 ohm 10X low impedance probe to look at the waveform of the oscillator. Try varying lengths of probe ground lead up to 1 foot while adjusting the frequency of the square-wave generator from 5 to 50 MHz.

Questions:

3.1.1 How sensitive are the displayed waveforms to the ground lead length of the high and low impedance probes?

Answer: Because of the reactive nature of the input imped-ance of the high impedance probe, it is much more sensitive to probe ground lead length and its attendant inductance. In general, the effects noted will include overshoot, ringing, amplitude peaking at certain frequencies, and loss of high frequency response in the measurement. The next chapter will deal with this subject in detail.

Experiment 3.2: Electrostatic Charge

1.   Set the scope vertical sensitivity to about 10 volts/division and the sweep to about 1 second/division. Connect a 10X high impedance probe to a vertical input of the scope. While holding the probe tip with two fingers, without touching any grounded metal, scuff one foot back and forth on the floor. Note any vertical movement of the trace as it moves across the screen.
2.   Repeat step 1 using a 10X low impedance probe.

Questions:

3.2.1   What mechanism causes the vertical movement of the trace as the foot is moved?

Answer: Electrostatic charge is generated on the bottom of the foot as the foot is moved on the floor, especially on rugs in dry weather. Only a few microamps of current flowing from the fingers to and from the scope probe are generated in this manner.   Since the probe input resistance is 10 megohms, this results in several volts being generated at the probe tip.

3.2.2   How does the vertical displacement of the trace compare between the high and low impedance probes? Why?

Answer: The low impedance probe shows much less vertical displacement of the scope trace. This is because the same current is generated into the probe tip by the motion of scuffing a foot. However, since the low impedance probe has an input impedance of only 500 ohms typically as compared to 10 megohms for the high impedance probe, much less voltage is produced at the probe tip.

PROBLEMS

3.1.   A 10 pF capacitor can have surprisingly low impedance at frequencies that are not particularly high by today's standards. What is the impedance of 10 pF of capacitance, a typical input capacitance for a passive high impedance probe, at 100 MHz? What affect does the series inductance of 12 inches of wire, 240 nH have on the total input impedance including the inductance?

Answer: A 10 pF capacitor at 100 MHz has only about 160 ohms of capacitive reactance. When placed in series with 12 inches of wire having 240 nh of inductance a series tuned circuit results. The resonant frequency is very close to 100 MHz and the resultant input impedance of the L/C series circuit becomes very close to the resistance of the connecting wires plus any resistive loss in the capacitor itself. The input impedance may approach a few ohms or less.

3.2.   Assuming that the impedance looking into a 50 ohm coaxial cable properly matched at the far end is 50 ohms over a wide frequency range and that the scope input matches the cable characteristic impedance, what is the input impedance of a 10X low impedance probe constructed using a 50 ohm coaxial cable and a series 450 ohm resistor that has 1 pF of capacitance between its end caps? Above what frequency does the input impedance reach its minimum. What is that minimum?

Answer: The 450 ohm resistor can be modeled as 450 ohms in parallel with 1 pF. Thus the input impedance is given by:

$$Z_{in} = \frac{R_{in} \cdot 1/(j \cdot 2\pi \cdot f \cdot C_{in})}{R_{in} + 1/(j \cdot 2\pi \cdot f \cdot C_{in})} + 50 \; ohms$$

where: R_{in} = 450 ohms, and
       C_{in} = 1 pF.

This equation can be rearranged as follows:

$$Z_{in} = \frac{R_{in}}{1 + j \cdot 2\pi \cdot f \cdot R_{in} \cdot C_{in}} + 50 \; ohms.$$

The frequency at which the reactance of C_{in}, 1 pF, equals the value of R_{in}, 450 ohms, is 354 MHz. Above that frequency, the input impedance of the probe begins to fall and the above equation reduces to:

$$Z_{in} = \frac{-j}{2\pi \cdot f \cdot C_{in}} + 50 \; ohms.$$

Above the frequency where the reactance of C_{in} equals 50 ohms, the probe input impedance rapidly approaches its minimum, 50 ohms. That frequency is about 3200 MHz.

4

Probe Ground Lead Effects, Resonance and Induction

INTRODUCTION—THE UBIQUITOUS PIGTAIL

In order to develop an understanding of high frequency measurement techniques, a good place to start is with the use of oscilloscope probes and sources of error that can occur from their use. Specifically, the ground lead used with many scope probes is a major source of error, especially when used with the common 10X high impedance passive probe. The error than can result is caused by two effects: probe resonance and common impedance coupling. As the reader will see, amplitude errors attributable to these effects can become extreme and severe wave shape distortion can be introduced. In this chapter, each of the two effects will be discussed in detail.

LEAD INDUCTANCE

Any conducting path contributes inductance to a circuit. The subject of partial inductances has been thoroughly discussed in the literature,[1]

---

1.   Ruehli, IBM Journal of Research and Development, Sept. 1972, pp. 470-481.

but for the purposes of this chapter an average figure of 20 nH per inch for small wires similar to scope probe ground leads will be used. The assumption that must be made is that the return path for the current is far enough from the wire segment under consideration so as to have negligible partial mutual inductive coupling to it. The magnetic field of a current in a conductor falls off rapidly with distance so a few inches are generally adequate for the types of circuits being considered here for measurement purposes.

As discussed in Chapter 2, at almost any frequency above the audio band, wires should be considered as inductors since the magnitude of their inductive reactance far exceeds the magnitude of their resistance. In the measurement context, this is a problem because the inductance of a probe ground lead can combine with any capacitive component of the probe input impedance to cause resonance, filtering of the signal, and subsequently, significant errors. For example, the 24 gauge wire discussed in Chapter 2 has a Q (X_L/R) of over 70 at 5 MHz. when considered as an inductor. The potential for forming a high Q resonant circuit should be obvious. The ground lead inductance can also be a source of common impedance errors. The remainder of this chapter is devoted to a discussion of ground lead resonance and common impedance effects.

LEAD INDUCTANCE AND PROBE RESPONSE

Equivalent Circuit

The equivalent circuit of a probe, its ground lead, and the measured circuit is shown in Figure 4.1. The probe input impedance, Z_p, is effectively in series with three inductances to form the load on the source composed of the Thevenin source voltage, V_{so} and the impedance of the source, Z_{so}. Inductance L_1 is the probe tip inductance, usually equivalent to about 3 inches of lead length, L_2 is the ground lead inductance, typically about 6 inches, and L_3 is the inductance of the conductors in the measurement circuit such as paths on a printed wiring board. For the purpose of the discussion of probe resonance, inductances L_1, L_2, and L_3 will be lumped into a single inductor L_g. L_g is usually dominated by L_2 for many measurements.

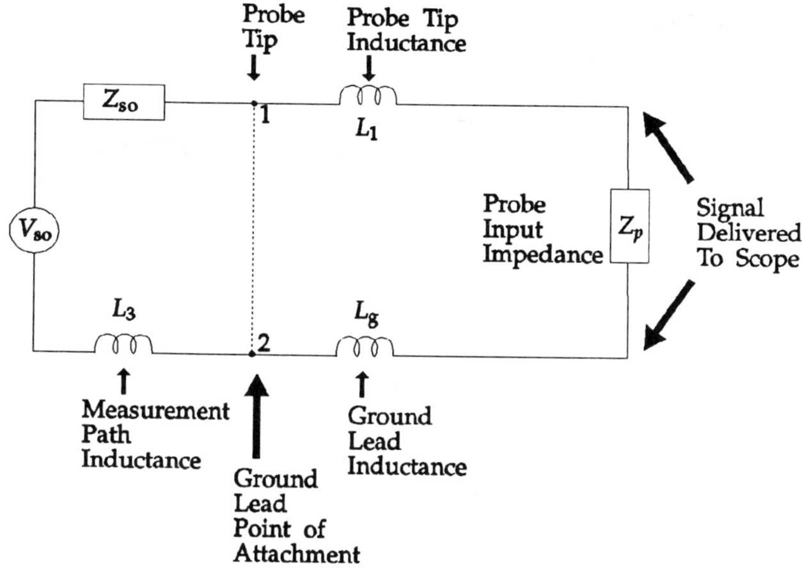

Figure 4.1 Measurement equivalent circuit.

For most measurements where a probe ground lead is used, it is difficult to reduce the path length forming L_g to less than 12 inches. A 6 inch ground lead, 3 inch probe tip, and 3 inches of signal path are not unusually long.

The voltage developed across the probe input impedance, Z_p, is the signal delivered by the probe to the scope for display. For the purposes of this discussion, it is assumed that the probe is well designed and compensated so that the signal delivered to the scope input is a faithful reproduction of the voltage appearing across Z_p, possibly reduced by a scale factor.

Probe Resonance

For probes where the input impedance, Z_p, is capacitive, such as a 10X high impedance passive probe, a potential for resonance exists. Figure 4.2 shows such a case using L_g to represent the total circuit inductance.

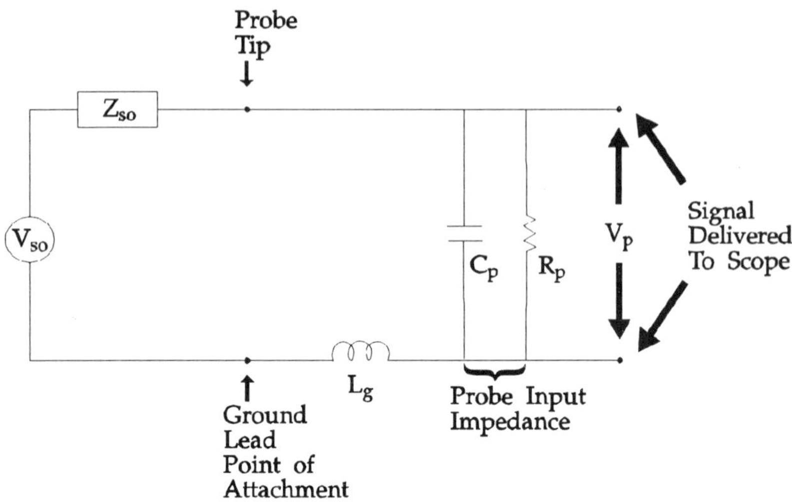

Figure 4.2 Measurement using high impedance 10X passive probe.

The probe input impedance is the parallel combination of C_p and R_p, typically 10 pF and 10 megohms. The circuit is damped only by the relatively small series wire resistance and the 10 megohm input resistance of the probe. As a result, a high Q tuned circuit is formed composed of C_p, R_p, and L_g.

Although the source impedance may provide some damping, in many measurements the resistive component of the source impedance is very low. Measurement of ground noise potential across a printed wiring board path is such an example. In this case, the source impedance is mostly inductive (it becomes part of L_g) and contributes to the problem.

Figure 4.3 shows a graph of probe input voltage, V_p, versus the source voltage, V_{so}, for a 10X passive probe with a 10 megohm/10 pF input impedance where the source impedance has an insignificant resistive component and L_g is composed of 12 inches of path length. The ordinate scale has a range of 0 to 40 dB (V_p/V_{so}) and a range of 20 to 200 MHz is covered by the abscissa.

Ideally, V_p/V_{so} should be unity, but the error exceeds 3 dB above about 60 MHz. At resonance, about 100 MHz, a significant gain peak occurs due to the high Q of the resonance. The Q can approach 100 since the only damping in the circuit is provided by the 10 megohm input resistance. Above the 100 MHz resonance, V_p is decreasing at 40 dB/decade due to the LC two pole filter formed by L_g and C_p. Frequency components of signals much above 100 MHz are lost.

The time domain step response of a circuit with the frequency response of Figure 4.3 exhibits severe overshoot and ringing. Such a response would be noticeable on signals whose wave shape is known or at least suspected, but on a measurement of a noise voltage of unknown wave shape, it may go undetected.

For some measurements, the signal or noise source output impedance will have a significant resistive component. Examples of such sources would include logic gates and laboratory signal generators. For these sources, the resistive component of the output impedance can provide significant damping of the resonant circuit. Unfortunately, unacceptable errors usually still exist in the measurement.

Figure 4.4 shows the frequency response for the same measurement with a high impedance 10X passive probe as in Figure 4.3 except that the source impedance of the signal is resistive and has a value of 50 ohms. The ordinate scale has been changed to +10 to −10 dB to more effectively display the response. Several important characteristics should be pointed out. First, the resonance has a much reduced Q, about 3, and the resultant gain peak is *only* 10 dB. Of

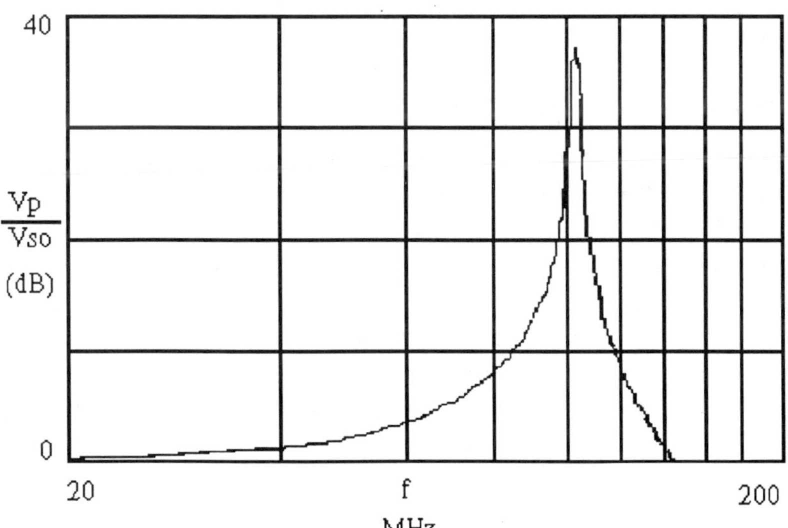

Figure 4.3 Probe response to a voltage source.

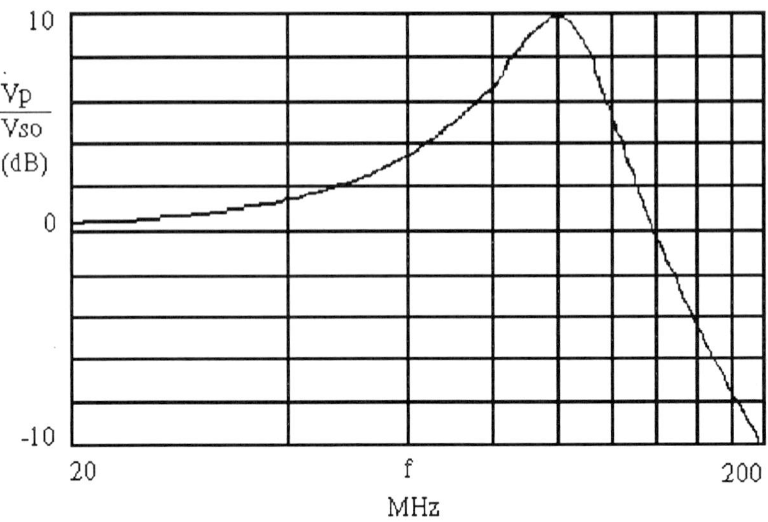

Figure 4.4 Probe response to a 50 ohm source.

course this is still a factor of 3, not exactly an accurate measurement. The reduced Q will also reduce the overshoot and ringing in the step response.

Note that the error is still about 3 dB above 60 MHz and that energy above 100 MHz is still being filtered out at 40 dB/decade as frequency increases.

There exists a simple way to minimize the degradation in probe response caused by resonance and this is implied by Figure 4.4. Consider the effect of adding 100 ohms of resistance in series with the probe tip of a 10X passive high impedance probe. At frequencies below a few tens of MHz the additional series impedance will not measurably affect the displayed waveform. However, at higher frequencies the extra series resistance will dampen the resonance and improve the measurement accuracy.

Figure 4.5 shows a graph of the measurement of Figure 4.4 with an added 100 ohm resistor in series with the probe tip. As in Figure 4.4, Figure 4.5 shows probe input voltage, V_p, versus the source voltage, V_{so}, for a 10X passive probe with a 10 megohm/10 pF input impedance and a source resistance of 50 ohms. L_g, the inductance of the measurement circuit, is composed of 12 inches of path length. The ordinate scale has a range of −10 to +10 dB (V_p/V_{so}), and a range of 20 to 200 MHz is covered by the abscissa.

With the added 100 ohm resistor, measurement error does not exceed 3 dB until about 120 MHz. The accurate measurement bandwidth has been increased a whole octave. Measurement response above 100 MHz is still falling off at 40 dB/decade. Signal energy above resonance can not be recovered by this method so it is still a good idea to minimize probe ground lead length.

The use of series resistance should be used to improve measurement accuracy when the minimum practical length ground lead is used. For probe ground leads of up to 4 to 6 inches and where a measurement bandwidth of 150 MHz or less is adequate, series resistance at the probe tip works well with standard 10X high impedance passive probes.

The value of the series resistance is a compromise between adequate damping and bandwidth. If a little resistance is good, a lot is not necessarily better. When the *total* of the source resistance plus the probe tip series resistance exceeds 150 to 200 ohms, bandwidth will be reduced by the RC filter composed of the probe input capacitance and the source plus tip resistance for most common 10X

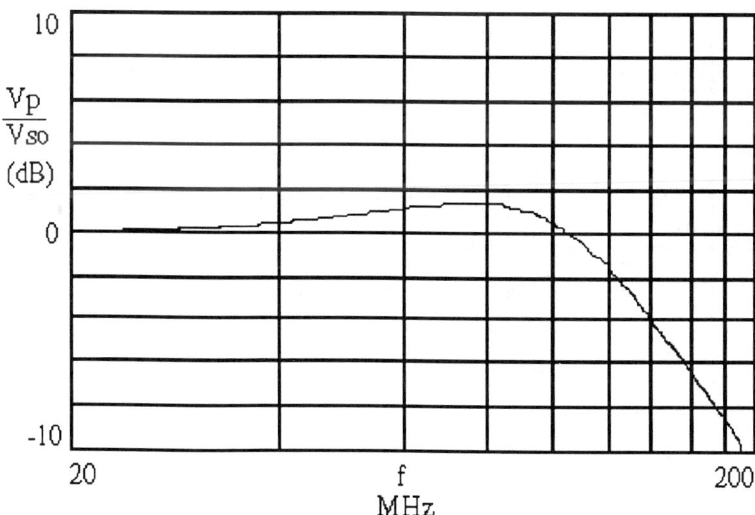

Figure 4.5 Probe response to a 50 ohm source with a series 100 ohm resistor.

high impedance passive probes. Figure 4.6 shows the same measurement as Figure 4.5 except that the 100 ohm probe tip resistor has been increased from 100 ohms to 300 ohms. The excessive RC rolloff in frequency response of the probe is evident.

Theoretically, one could adjust the added series tip resistor to an individual probe or better yet to a total measurement circuit to obtain the optimum response. In practice, this adjustment is usually not worth the effort. Most high frequency (above a few tens of MHz) signals or noise sources have source resistances that range from near zero to about 100 ohms or so. Thus, for most cases an added series resistor of 100 ohms at the probe tip will yield a total measurement circuit resistance of 100 to 200 ohms. Over this range, reasonable measurement accuracy will be obtained with neither a high Q resonance or excessive RC rolloff occurring.

Probe Input Impedance

Probe input impedance can affect circuit operation as well as measurement accuracy and for some probe types can be much lower than is normally assumed. Take for instance the 10X passive high imped-

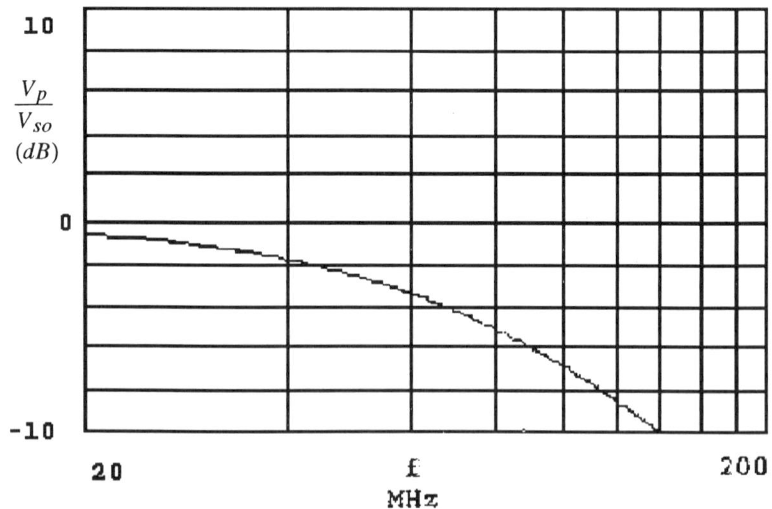

Figure 4.6 Probe response to a 50 ohm source with a series 300 ohm resistor.

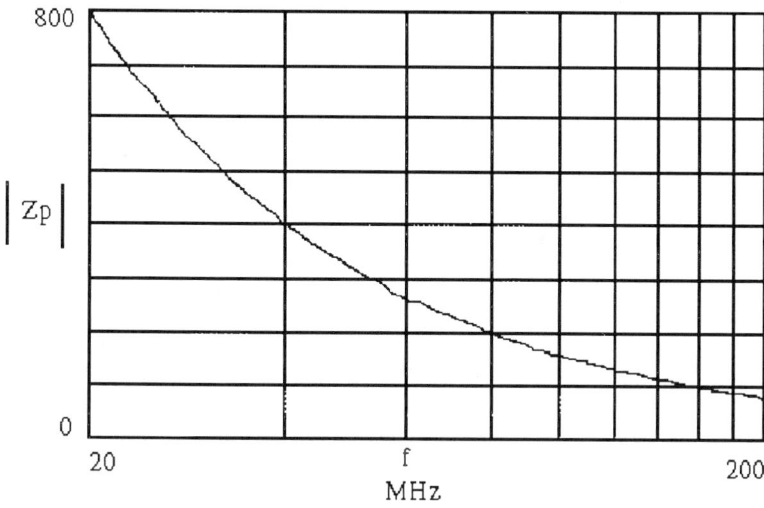

Figure 4.7 Probe input impedance.

ance probe discussed above. The name of the probe includes the words "high impedance," however its input impedance can become quite low over a large range of measurement frequencies as shown in Figure 4.7.

Figure 4.7 shows the magnitude of the impedance of 10 megohms in parallel with 10 pF, a typical input impedance of a 10X passive probe, versus frequency over the range of 20 to 200 MHz. The input impedance starts out at less than 800 ohms at 20 MHz, lowering to less than 100 ohms at 200 MHz. This impedance is not all that high and, in fact, above about 30 MHz the 500 ohm 10X low impedance passive probe has a higher input impedance.

Probe input impedances on the order of that shown in Figure 4.7 can affect circuit operation. Unfortunately, at least for the 10X high impedance passive probe, the problem is even worse than that shown in Figure 4.7. The inductance of the measurement path, mostly ground lead inductance, can lower input impedance further at the resonant frequency as shown in Figure 4.8.

Figure 4.8 shows that the input impedance of a 10X high imped-ance passive probe can drop to near zero at the resonant frequency of the lead inductance and the probe input capacitance. The network

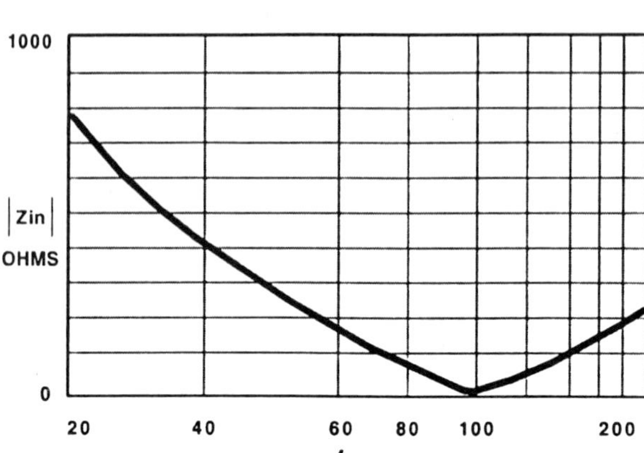

Figure 4.8 Probe input impedance including lead inductance.

used to generate the curve in Figure 4.8 is composed of a 240 nh inductor (12 inch path length) in series with the parallel combination of a 10 pF and 10 megohms (a typical 10X high impedance passive probe input impedance).

In effect, connecting a high impedance passive probe with a ground lead to a circuit imposes a notch filter on the circuit. If the notch happens to coincide with the frequency spectrum of some noise source that is causing a malfunction, it may be filtered out by the connection of the probe, causing the circuit to function properly.

Many engineers have experienced the effect of connecting a scope to a malfunctioning circuit only to have the circuit start working correctly and, under the pressures of a tight schedule, it is easy to believe that some engineers have been tempted to leave a scope probe in the equipment being tested "for the convenience of field service personnel." Actually, there are a few effects that can cause circuits to start working when probes are connected, but probe resonance is the most likely (another mechanism will be discussed later in this chapter).

There is a simple test circuit that is easily built and can determine if probe resonance is affecting circuit operation. Called a "tuned

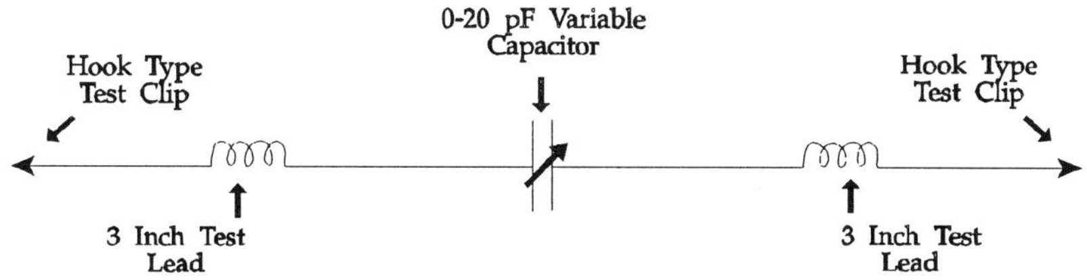

Figure 4.9 Tuned probe tester.

probe tester," it is simply a test clip that is a few inches in length, with a small variable capacitor covering the range of a few pF to few tens of pF connected in series. The tuned probe tester is connected to the circuit being measured instead of the probe and the variable capacitor is adjusted. If circuit operation is improved the same way as with the probe connected, then probe resonance was the likely cause of probe connection resulting in proper circuit operation. Figure 4.9 illustrates a tuned probe tester.

Just as the addition of a 100 ohm resistor to the probe tip improved the frequency response of the 10X high impedance passive probe by lowering the Q of the resonance, the input impedance picture is also improved. Figure 4.10 shows the input impedance of a 10X high impedance passive probe including 12 inches of path inductance *and* a 100 ohm resistor. As one might expect, the input impedance at resonance is raised from near zero to about 100 ohms. Although not as high as one would like, an input impedance of 100 ohms at resonance results in much less circuit loading.

PROBE TYPES WITH IMPROVED RESPONSE

Low Impedance Passive Probes

Unlike the 10X high impedance passive probe, some probe types have well behaved frequency response and input impedances over a wide frequency range, well above the 100 MHz resonance discussed earlier in this chapter. In general, for both response and input impedance, these probes are less sensitive to ground lead inductance than standard 10X high impedance passive probes. Two examples of such probes are low impedance passive probes and active high impedance probes (1 pF in parallel with 10 megohms).

Low impedance passive probes, discussed in Chapter 3, have a mostly resistive input impedance that precludes high Q resonance. As the input frequency increases, the inductance of the ground lead and the rest of the measurement path becomes comparable to the input resistance of the probe. For a 12 inch path length with an inductance of 240 nH and a probe with a 500 ohm input impedance, the inductive reactance of the measurement path equals the input resistance at about 330 MHz. Above this frequency the response of the probe falls off at 20 dB/decade due to the L/R single pole low pass filter response of the input resistance and measurement path inductance. Since the roll off is single pole, there is no resonant peaking. Above 200 to 300 MHz, the scope input capacitance can further roll off probe frequency response, as described in Chapter 3. Scope input capacitance is usually lowered to a few pF if a 50 ohm input on the scope is available and used.

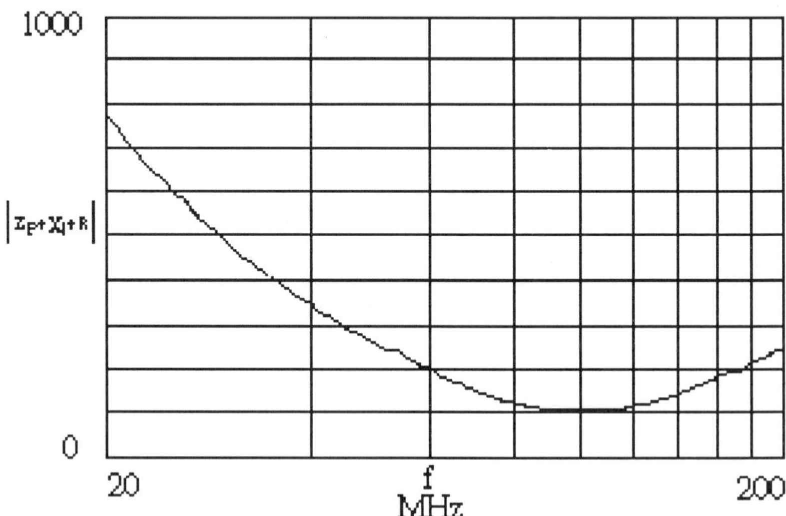

Figure 4.10 Probe input impedance including lead inductance and a 100 ohm resistor.

Low impedance passive probes are readily available. Commercial low impedance passive probes costing a few hundred dollars have bandwidths of 1 GHz and higher. With reasonable care, one can build a 10X 500 ohm low impedance passive probe having flat response to 500 MHz or more for less than $20 using readily available parts. See Chapter 3 for more details on the construction and performance of this type of probe.

Oscilloscope probes take a lot of abuse in typical laboratory settings. The abuses range from electrical overstress to physical damage such as being run over by a scope cart or stepped upon. As an added benefit, low impedance passive probes are generally rugged, both electrically and mechanically. Much less care is needed in handling these probes.

High Impedance Active Probes

The low input capacitance of high impedance active probes moves the resonance to a much higher frequency. Since resonant frequency changes inversely with the square root of capacitance, lowering the probe input capacitance from 10 pF (high impedance passive probe) to 1 pF (high impedance active probe) raises the resonant frequency of the probe input capacitance and the measurement path inductance by approximately a factor of 3. For the case of a 6 inch ground lead and a total measurement path length of 12 inches, the resonant frequency is moved from about 100 MHz to over 300 MHz. Often this is high enough to allow accurate measurements. Probe resonance could still be a problem though at the higher frequencies.

High impedance active probes can be expensive and fragile. The advantages of low circuit loading and high resonant frequency often offset the disadvantages of these probes. For measurement of high Q, high frequency resonant circuits, a probe of this type is necessary.

TELL TALE SIGNS OF PROBE RESONANCE

Probe resonance manifests itself in several ways that indicate to the user what is happening. One is overshoot and ringing. Any waveform that contains overshoot and ringing should be suspected and steps

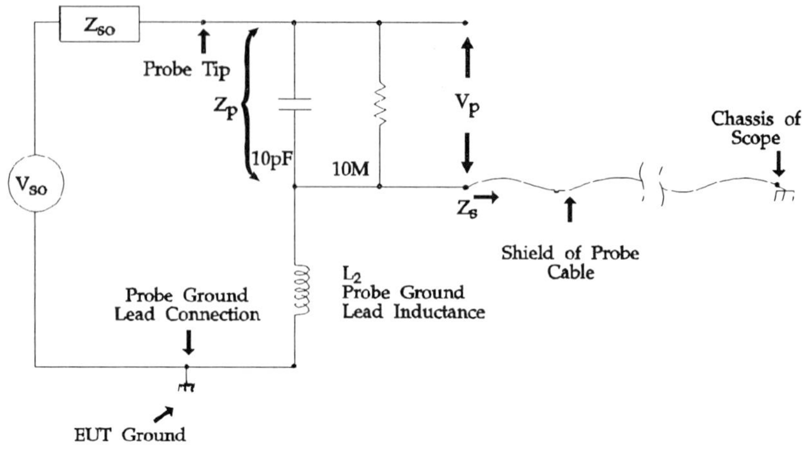

Figure 4.11 Probe resonance equivalent circuit.

should be taken to determine if it is the probe or if the effects really have arisen in the circuit. The use of a probe ground lead of any substantial length, greater than an inch or so, increases the suspicion that it is the probe response and not the circuit. Wiggly scope patterns are another indicator of probe resonance.

Wiggly Scope Patterns

Wiggly scope patterns are those that seem to change shape as the probe cable is moved or handled. When this effect is present, the amount of overshoot and the frequency of ringing are changed slightly as the probe and cable are handled. The presence of this effect should create doubt about the validity of the waveform. What causes it?

The cause of wiggly scope patterns can be understood with reference to Figure 4.11. In this figure, the probe input impedance, Zp, forms a series resonant circuit with the ground lead inductance, Lg. The voltage appearing across the probe input impedance, Vp, is delivered to the scope and is displayed, usually it is divided by a scale factor, with the probe's stated accuracy. Note that the probe cable shield is

connected to the center of the series resonant circuit and the impedance looking into the shield, Zs, is basically in parallel with the ground
lead inductance.

The impedance looking into the probe cable shield is a function
of many variables. These include parasitic capacitance to ground,
shield inductance, and radiation efficiency of the shield as an antenna. This impedance can vary from less than 100 ohms upward to
1000 ohms or more with varying real and imaginary components.
Whatever the impedance, it is certainly changed when the cable is
handled or moved and this changing impedance is across the inductance of the tuned circuit shown in Figure 4.11. The resonant frequency and the Q of the tuned circuit composed of the probe, its
ground lead, and the cable shield impedance vary as the cable is
handled or moved and thus the overshoot and frequency of ringing
is changed as well.

The presence of wiggly scope patterns as the probe cable is
handled is a sure sign that probe resonance is affecting the measurement. When this happens, steps such as reducing the length of the
probe ground lead or using a different probe type should be taken to
minimize resonance.

GROUND LEAD COMMON IMPEDANCE INDUCED ERRORS

Unfortunately, probe resonance is not the only effect of the ground
lead. Noise currents caused by sources other than the signal being
measured can flow from ground in the Equipment Under Test (EUT)
at the point where the probe is connected through the probe ground
lead and onto the probe cable shield. The sources of this noise current
can be internal to the EUT or external to it, a nearby electrostatic
discharge event for example. The magnitudes of the errors generated
by this noise current can easily exceed those of the signals being
measured.

Figure 4.12 shows a typical measurement configuration and the
path that the noise current takes. A high frequency, a few MHz or
more, potential difference between the scope chassis and the local
ground in the EUT causes a high frequency noise current to flow
from the EUT ground on the scope probe ground lead and onto the

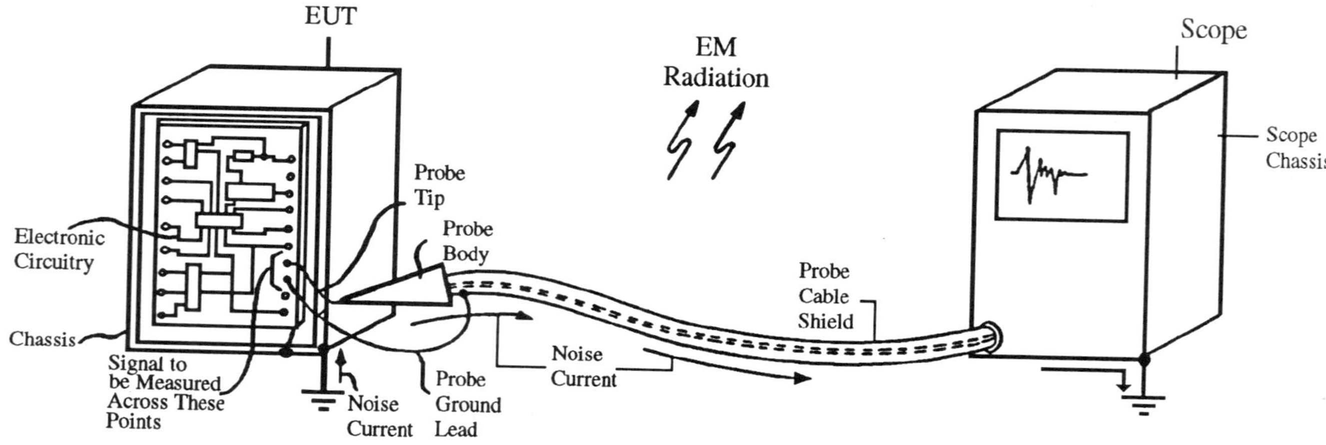

Figure 4.12 Typical measurement configuration with ground noise currents.

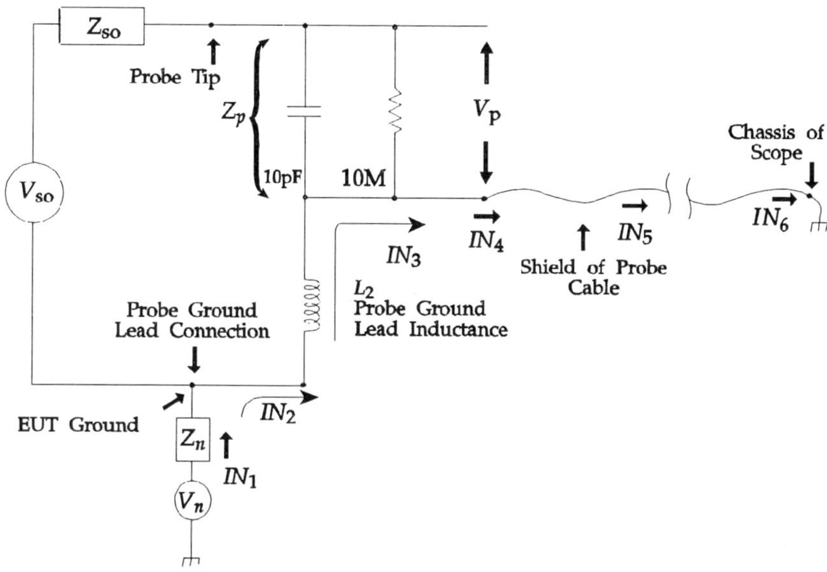

Figure 4.13 Simplified equivalent circuit of typical measurement configuration with ground noise currents.

shield of the probe cable. Some of the current is radiated from the probe shield as an antenna and some is conducted into the scope chassis.

The mechanism by which this noise current generates a measurement error can be seen by reference to Figure 4.13, which shows a simplified equivalent circuit of the configuration of Figure 4.12.

V_n and Z_n, the Thevenin equivalent noise voltage and source impedance respectively, cause the local signal ground in the EUT to assume a potential with respect to the scope ground and drives a noise current, I_n, though the ground lead inductance, L_g, down the shield of the probe cable and into the scope. The magnitude of this current at the several locations shown, In_1 through In_6, can vary because of standing waves, radiation from the cable shield, and parasitic capacitive coupling to nearby conductors. As the noise current flows, it generates a voltage drop across the probe ground lead and the shield of the probe cable. Noise current IN_3 causes the voltage drop across

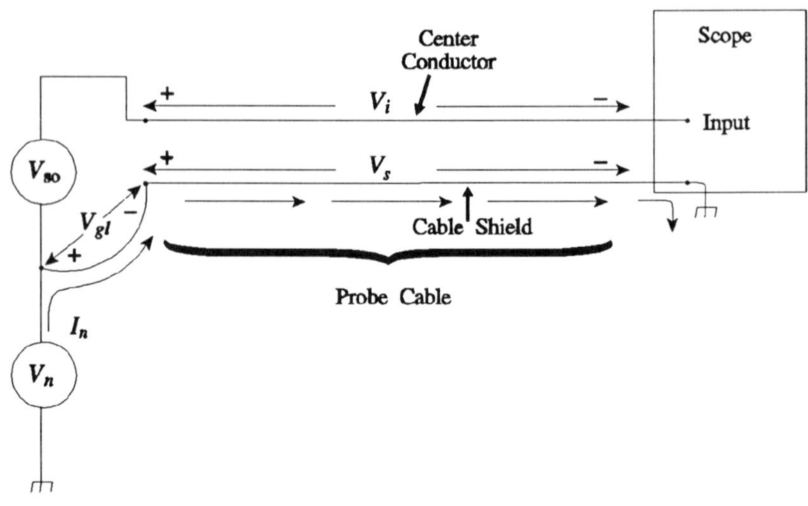

Figure 4.14 Simplified measurement circuit

L_g and this voltage is in series with the source voltage, V_{SO}, that is being measured.

The noise current flowing on the shield of the probe cable also causes a voltage drop across the shield. Its effect can be understood by reference to Figure 4.14. For ease of understanding, the circuit of Figure 4.14 has been simplified to show only the noise source, V_n, the signal source to be measured, V_{SO}, the ground lead, probe cable, and the oscilloscope. The probe cable center conductor and shield are shown as two parallel wires to clearly show where the induced voltages are.

In Figure 4.14, noise source, V_n, drives a noise current into the probe ground lead and cable shield. In the process, V_{gl}, the impressed noise voltage across the ground lead and V_s, the impressed noise voltage across the shield are generated. In addition to the voltage drop across the cable shield, the current in the shield also induces a voltage into the center conductor, V_i, with the same polarity as the voltage across the shield whose magnitude is given by Equation 2.22 as derived in Chapter 2:

$$V_i/V_s = \frac{j\omega L_s}{R_s + j\omega L_s} = \frac{j\omega}{j\omega + R_s/L_s} = \left[\frac{1}{1 + \dfrac{R_s}{j\omega L_s}}\right] \qquad (2.22)$$

where: V_i is the induced voltage into the center conductor,
      V_s is the voltage across the shield due to the noise current,
      R_s is the resistance of the shield, and
      L_s is the inductance of the shield.

The corner frequency of Equation 2.22, $(1/2\pi) \cdot (R_s/L_s)$, is on the order of 1 to 100 kHz for most shielded cables. Thus at frequencies above 1 MHz, the induced center conductor voltage V_i is for all purposes equal to the shield voltage V_s.

By adding up the voltages around the measurement loop, one can determine the signal across the scope input. The total of these voltages is:

$$V_{scope} = V_s + V_{gl} + V_{so} - V_i \qquad (4.1)$$

where: V_s is the voltage impressed across the cable shield,
      V_{gl} is the voltage impressed across the probe ground lead,
      V_{so} is the intended signal source to be measured, and
      V_i is the voltage induced into the cable center conductor.

The center conductor voltage, V_i, is induced with the same polarity as the shield noise voltage but is negative in Equation 4.1 because as the voltages are summed around the loop its polarity is opposite to the shield voltage as shown in Figure 4.14. But, since the magnitude of V_i is about the same as V_s, Equation 4.1 reduces to:

$$V_{scope} = V_{sig} + V_{gl} \qquad (4.2)$$

where: V_{gl} is the voltage impressed across the probe ground lead, and
      V_{sig} is the intended signal source to be measured.

It is seen from Equation 4.2 that the noise voltage across the ground lead is directly in series with the intended signal voltage. Since the probe cable does not have a perfect shield with uniform current, some of the shield current will also generate an error voltage. Usually this error voltage is 20 to 40 dB down from the error caused by even a short ground lead of an inch or two.

The error voltage caused by the voltage across the ground lead can be significant. Table 4.1 shows the $L \cdot di/dt$ voltage drop across 6 inches of wire, 120 nH of inductance, for a 10 ma change in current over the listed time.

Table 4.1 Inductive drop across 6 inches of 24 gauge wire.

Risetime	Induced Voltage
1 ns	1200 mv
2 ns	600 mv
3 ns	400 mv
4 ns	300 mv
5 ns	240 mv
6 ns	200 mv
10 ns	120 mv

The voltages listed in Table 4.1 are large enough to introduce significant error into the measurements. A 10 ma change in 1 nanosecond produces a 1.2 volt inductive drop, and even that is small compared to ESD induced voltages, switching power supply noise, and even some sources of logic noise. Needless to say, an attempt to measure logic noise margins in the face of a 1.2 volt measurement error is not likely to succeed.

THE NULL EXPERIMENT

Given that large measurement errors such as described above can happen, how can the error itself be measured? An experiment to make this measurement could be called a Null Experiment. A null experiment is a measurement that should have an obvious outcome, usually zero. This concept is an important one and amounts to devising an experiment to validate a measurement. Null experiments should always be done before basing any decisions on the outcome of a measurement, such as changing a circuit design to improve the noise margin. Maybe the noise margin is already adequate.

For a scope probe, one null experiment would be to eliminate the signal voltage to be measured and just measure the ground lead voltage. This can be accomplished by simply shorting the probe ground lead to the probe tip and connecting them both to the point on the EUT ground where the ground lead would be connected for the intended measurement. The signal displayed by the scope under this condition is the margin of error for the measurement. It is not unusual to find voltages from logic noise of hundreds of millivolts. Other sources of ground lead induced voltage can produce a reading of many volts peak from the probe null experiment. Table 4.2 lists some sources of error and the voltages that can be induced per inch in a scope ground lead, or any other lead for that matter.

Table 4.2 Typical voltages induced in a conductor by several sources.

Source	Induced Peak Voltage per Inch
Digital logic	200 mv
Switching power supplies	5 v
Electrostatic discharge	100 v
Radio frequency transmitters	10 mv

The voltages listed in Table 4.2 are not maximum values but represent relatively common values than can be encountered. ESD is among the most potent sources of interference to measurements and is capable of introducing extreme error into a measurement. The situation is made worse by the fact that ESD is not considered in many measurements that do not directly measure an ESD event or its result. Measurements in an ESD environment will be discussed at some length in Chapter 9.

EXAMPLE:

4.1 Consider the measurement of noise transients on a 5 volt supply using a storage or digital scope, a common measurement when equipment is intermittently malfunctioning. This measurement is almost a self-fulfilling search for noise pulses. It is self-fulfilling because a noise transient is almost always found. A typical transient found will be about 3 volts peak-to-peak with 20 ns oscillations. If the scope probe tip is moved from the probe ground lead connection to the EUT, generally the noise transient will remain the same, just without a 5 volt offset. Often, the cause is an ESD event in the same room, usually so small as to be unnoticed, on the order of several hundred volts. The ESD event generates fields that cause current to flow between the EUT and the scope over the scope probe shield and ground lead. The transient displayed on the scope is mostly the voltage induced across the inductance of the ground lead. The scope probe and ground lead also form a loop antenna, but its pickup is usually much smaller than the voltage induced as described above.

There are many sources of noise that will induce ground lead errors into scope probe measurements of which the example above is just one. It is therefore necessary to use the shorted probe null experiment to help verify any measurement involving frequencies much above the audio band.

GROUND LOADING

Attaching the ground lead of a scope probe can also affect equipment operation, all by itself. The same current that generates a voltage drop across the probe ground lead must also cause a redistribution of ground currents in the EUT. Ground noise voltages are thus altered in the EUT and operation may be affected. This effect can be termed "ground loading." How can one determine if probe resonance or ground loading is affecting equipment operation?

One way is to replace the probe and its ground lead with a tuned probe tester as discussed earlier in this chapter. The tuned probe is used to replace the scope probe and ground lead connected to the EUT and the variable capacitor is tuned to vary the resonant frequency. If the equipment is affected by adjusting the capacitor in the same way as the probe connection did, it is likely that probe resonance and not ground loading was affecting the circuit.

USE OF FERRITES ON PROBES

The type of ferrite cores used in electromagnetic interference control can also be used to improve the measurement accuracy of scope probes by reducing the common mode ground currents on the cable shield. Typically, passing a conductor through a ferrite core of this type will effectively introduce the series impedance of an inductor in parallel with a resistor in the conductor. A popular core generates a series impedance of about 100 ohms, mostly resistive, at about 100 MHz. By placing the core around the probe cable, ground currents on the cable shield are reduced, thus reducing the induced drop across the ground lead and ground loading of the EUT as well.

The position of the ferrite core on the probe cable is important. For convenience, one would like to place the core at the scope end. This would make the probe lighter and easier to handle, however the core effectiveness would be reduced substantially by locating the core at the probe end of the cable. Even if the core could produce an equivalent common mode open circuit at the connection of the probe cable to the scope, the probe cable shield could be a good enough antenna to allow a substantial current to flow in the probe ground lead that is subsequently radiated from the cable shield. Substantial current could also be coupled off the cable shield through parasitic capacitance between the probe cable shield and nearby metal objects.

The addition of the ferrite core to the probe cable will not affect the signal being measured. This is because the intended signal passes through the core on the center conductor and returns through the core on the shield resulting in no net signal current flowing through the core.

Substantial reduction of noise currents on the probe shield can be achieved with the use of ferrite cores this way. At some frequencies, the impedance looking into the probe ground lead and down the shield will be relatively low, a few hundred ohms or less, and significant noise currents can flow in the ground lead. The exact frequencies where this happens are a function of cable placement, length, capacitance to ground, and countless other parasitics, but the exact frequencies are not important. Broadband noise generated by most digital circuits will find those frequencies at which current can be driven into the ground lead and cable shield. Typically, the use of a ferrite core at the probe end of the cable can reduce the peak voltage drop across the probe ground lead by factor of two. This is often enough to allow a given measurement to be made. Some scope probe manufacturers market probes with the ferrite built into the probe body for this purpose.

MORE WIGGLY SCOPE PATTERNS

Just as with probe resonance, the changing impedance looking into the probe shield from the EUT as the cable is handled affects the noise current flowing and therefore the voltage drop across the ground lead. In this case, if the cable is handled while the probe is shorted by its ground lead and connected to the EUT ground, the results of this null experiment will change as the cable is moved. The addition of ferrite to the probe end of the cable will substantially reduce this effect in many cases.

For high impedance passive probes, probe resonance is strongly coupled to the voltage developed across the ground lead. This can be understood by reference to Figure 4.15 which shows the equivalent circuit of the shorted probe null experiment.

In the figure, the probe input capacitance and the inductance of the probe ground lead form a high Q parallel tuned circuit. The impedance of such a circuit to current flowing into the probe shield from the EUT is very high at resonance. Thus the voltage developed across the probe ground lead can be substantially higher than that calculated by multiplying the probe shield current by the inductive

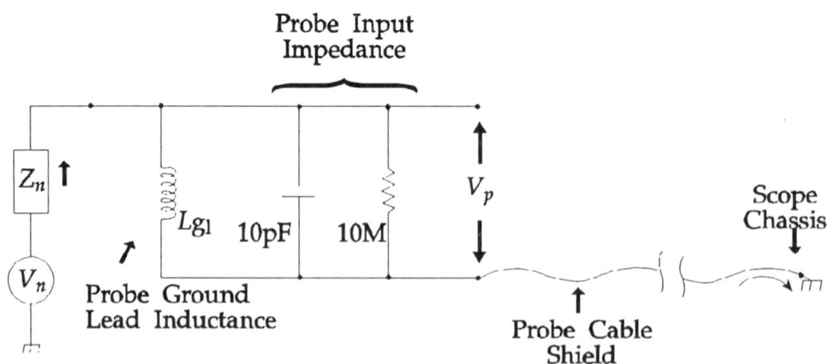

Figure 4.15 Equivalent circuit of null experiment.

reactance of the probe ground lead. This voltage is the probe input voltage, V_p, for this case.

Low impedance passive probes do not exhibit a strong resonance effect and therefore yield significantly lower values of voltage from the null experiment. A low impedance passive probe typically yields a null experiment output of one third or less than that of a high impedance passive probe with a ground lead of the same length on typical digital circuits. For this reason, low impedance passive probes generally are preferable to high impedance passive probes when making measurements in the presence of other noise sources such as digital logic or electrostatic discharge.

SUMMARY

In this chapter, sources of oscilloscope probe measurement error caused by the probe ground lead were discussed. These errors arise because of probe resonance and ground lead voltage drop caused by noise currents flowing from the EUT into the probe cable shield through the ground lead impedance. Errors from these two sources can exceed the magnitude of the signal being measured. Low imped-ance passive probes incur much less measurement error from the ground lead than do high impedance passive probes, because of the lack of a strong resonance below 500 MHz, and are preferred whenever possible.

The concept of a null experiment was developed. For a scope probe, the null experiment is to short the probe with its ground lead and connect it to the ground of the EUT. The displayed voltage on the scope represents the margin of error for subsequent measure-ments on that EUT. Scope displays of several volts peak resulting from the null experiment are possible. Low impedance passive probes generally yield lower null experiment voltages than do high impedance probes and should be used whenever possible.

Connection of a scope probe to equipment can cause the equip-ment operation to be affected, either in a positive or negative sense. This can either be caused by the low input impedance of a probe at resonance or by ground loading, because of the connection of the probe ground to the EUT. The use of a tuned probe tester, a 6 inch test lead with a series variable capacitor, can be used to identify probe resonance effects on the EUT.

Ferrites can be used to reduce the noise current flowing on the probe shield and ground wire. The ferrite core should be of the lossy type and located as close as possible to the probe end of the cable. Typically, a reduction of the ground lead voltage to about one half of its original value can be expected in many measurement situations.

LABORATORY DEMONSTRATIONS

Experiments

Equipment needed:

- 100 MHz or higher bandwidth scope (switchable 50 ohm input is desirable),
- 5-50 MHz square wave generator with rise/fall times less than 5 ns, such as described in Chapter 3, Figure 3.8,
- 10X passive high impedance probe,
- 10X 500 ohm low impedance probe (a homemade version constructed of a length of 50 ohm cable and a 450 ohm resistor at the end of the center conductor can be used), and
- several lengths of test leads with hook ends.

Experiment 4.1: Probe Resonance

1. Set the generator to about 5 MHz and connect it through a coaxial cable, terminated in 50 ohms, to the oscilloscope (use the 50 ohm input if available). Adjust the sweep speed so that two or three cycles are displayed on the screen and note the waveshape. The displayed waveform represents an accurate representation of the generator output waveform.
2. Repeat step 1 using a 10X high impedance passive probe but without using a ground lead. This can be done if the generator has a male BNC output connector by positioning the probe so that the tip contacts the center conductor of the BNC jack and the ground sleeve of the probe rests on the ground lip of the

BNC jack. Note the displayed waveform. This represents the best the probe can do without ground lead effects and should look similar to the waveform in step 1 except for DC loading if the probe is working correctly. This waveform will be compared to those in several steps below. If a 50 ohm BNC feedthrough termination is connected to the output of the generator and the probe is connected to the termination in the same way (no ground lead), the displayed waveshape would be identical to that seen in step 1 if the probe were ideal.

3. Connect a test clip between the probe ground lead and the ground lip of the generator BNC output so that the total length of the ground lead is about 10 to 12 inches. Tip the probe vertically so that the ground sleeve no longer contacts the BNC ground lip. Note the overshoot and ringing on the waveform. For comparison, short out the ground lead momentarily by lifting up and leaning the probe over so that the ground sleeve again contacts the BNC ground lip. Repeat this step with only the probe's ground lead in the circuit and note what happens to the waveform.

4. Repeat step 3 but vary the frequency from 25 to 50 MHz to find probe resonances. The 12 inch ground lead should lower the resonant frequency into this range. As the frequency is varied there will be discrete frequencies where the waveform peaks. At these frequencies there will be substantial waveshape and amplitude distortion on the displayed waveform.

5. Repeat steps 3 and 4 first with a 100 ohm resistor in series with the probe tip and then try 330 ohms. Clip the resistor leads to about 1 inch and push one end into the BNC jack center conductor. The other end can then be touched with the probe. For comparison, touch the probe tip to both sides of the resistor to see what effect the resistor has on the displayed waveshape. The use of the 100 ohm resistor substantially improves the probe's impulse response.

6. Repeat steps 3 and 4 using the 10X 500 ohm passive low impedance probe. Note that this probe is much less sensitive to ground lead effects than is the 10X high impedance passive probe.

Questions:

4.1.1   In step 1, a small glitch may occur about 10 to 20 ns after each transition. If the output of the generator is greater than 50 ohms the glitch will be in the opposite direction from the most recent transition (a positive glitch after a negative going edge). What causes this and how may it be reduced or eliminated?

Answer: In the one megohm input position, most scopes have between 10 and 30 pF of input capacitance. This capacitance causes a mistermination of the 50 ohm probe cable above the frequency where the reactance of the scope input capacitance becomes less than 50 ohms. A negative reflection is generated and reflected with no inversion from the probe tip end of the cable (if the generator output is lower than 50 ohms, an inversion will occur) and arrives back at the scope input 2 times the cable transmission time later. Many cables have a propagation speed of 1 ft/1.8 ns. Thus, a 5 foot cable will have a one way propagation delay of 9 ns and the glitch will show up 18 ns after each transition.

There are three ways to minimize this effect. One is to use a scope with a 50 ohm input resistance. These inputs typically have only a couple of pF of input capacitance, not enough to produce a noticeable effect for this experiment. Alternately, a 20 dB (10X) attenuator can be added at the scope input. The signal is attenuated by 20 dB (probe becomes 100X), but the reflection passes through the attenuator twice and is reduced by 40 dB. Finally, match the cable at the generator end so the reflection is absorbed there.

4.1.2   In step 3, the overshoot and ringing is less severe when the 10X high impedance passive probe ground lead is not extended with a test clip. For a 4 to 6 inch ground lead the overshoot on a 5 MHz square wave with 2 to 3 ns rise and fall times might be only about 30 percent of the signal amplitude with a 10 ns width and actually be a small feature on a 5 MHz square wave. Why is this overshoot important?

Answer: A possible answer to this question involves measurement of ground-to-ground or power to ground digital noise on a printed wiring board. For this type of measurement, the wave shape is less important than its maximum peak value. Reading the peak value at 30 percent more than the real amplitude could cause additional cost to be added to a printed wiring board design to fix a "non-problem."

4.1.3   In step 4, substantial amplitude peaking only occurred at discrete frequencies with the 10X high impedance passive probe. Would measurements on digital circuits be affected if the clock frequencies did not correspond to one of these discrete frequencies?

Answer: Most digital circuitry generates broadband noise. The probe will "tune into" the energy near its resonant frequency and display this energy with disproportionate amplitude.

4.1.4   How effective is the 100 ohm series resistor, used at the tip of a 10X high impedance passive probe, at reducing overshoot on the 5 MHz square wave?

Answer: In many instances, the use of a 100 ohm series tip resistor will remove most of the overshoot caused by a 4 inch ground lead. Remember, however, that the ground lead inductance still limits the upper bandwidth of the measurement.

4.1.5   Why is the low impedance passive probe less sensitive to the length of the ground lead? What effect does the ground lead have?

Answer: Its input impedance is mostly resistive, and so a high Q resonance cannot occur. The ground lead begins to affect probe performance when the reactance of the ground lead becomes comparable to the input impedance of the probe. For a 500 ohm probe and a 6 inch ground lead, the frequency where this occurs is about 600 MHz.

Experiment 4.2: Induced Ground Lead Voltage

1. Connect the ground of a high frequency square wave signal generator to a scope chassis with about 4 to 5 feet of test lead. This is very important, especially if the battery powered generator described in Chapter 3 is used, as the ground connection provides the return current path for the experiment. Using a 10X high impedance passive probe, connect the probe ground lead through a test lead extension, so that the ground lead is about 10 to 12 inches long, to the probe tip and connect the probe tip also to the output of the signal generator. If the generator has a BNC output it is convenient to put a 1 inch stub of #20 wire in the center pin receptacle and clip the extended probe ground lead and the probe tip to this wire. There is no connection to the BNC ground lip as the ground return is provided by the separate ground lead. Dress the probe cable to lie close to the ground lead. In this setup, the generator is driving a current down the probe ground lead, through the probe cable shield, into the scope chassis, and back on the ground return lead.

2. Vary the frequency of the square wave from 20 to 50 MHz. At discrete frequencies the waveform displayed on the scope will peak to values that represent almost as much as the peak amplitude available from the generator. For the battery powered generator described in Chapter 3, which has a 5 volt peak output, the scope may display an amplitude of 2 to 4 volts peak at those frequencies where peaking occurs. Peaking will be very sensitive to the exact placement of the probe cable and ground return lead. Remember, the displayed waveform is the voltage across the probe ground lead, a few inches of wire.

3. At the frequency of the highest peaking, put a ferrite core intended for electromagnetic interference suppression around the probe cable near the probe end. Typically, the peak value displayed will be reduced to about one half of the original value. Repeat at a frequency where peaking does not occur. This time the core does not affect the peak reading substantially.

4. Repeat steps 1 to 3 without extending the probe ground lead. The displayed waveforms will have substantially reduced peak values.

5. Repeat steps 1 to 4 using a 10X 500 ohm passive low impedance probe. For similar test conditions the displayed waveforms will

have much smaller peak values than with the 10X high impedance passive probe.

Questions:

4.2.1   Why do the displayed waveforms peak at certain frequencies?

Answer: The impedance of the loop formed by the probe cable and the ground return lead as seen by the signal generator varies over a wide range with frequency. At those frequencies where the impedance of the loop is low, the generator can drive more current and therefore more voltage is induced across the probe ground lead. Complicating this picture is the resonance of high impedance passive probes.

4.2.2   Why does the ferrite added to the probe cable help mostly at the frequencies where peaking occurs?

Answer: The ferrite adds a hundred or so ohms of resistance to the high frequency path. At the peaking frequencies, this added resistance substantially reduces current flow. At non-peaking frequencies, the impedance of the circuit is much higher and another hundred ohms or so makes little difference.

4.2.3   Why does the low impedance passive probe develop less voltage across its ground lead than the high impedance passive probe?

Answer: Probe resonance causes the impedance of the parallel combination of the probe ground lead and the input impedance of the probe to become high near its resonant frequency. The reactive nature of the probe input impedance can also affect the impedance of the complete circuit composed of the probe cable and ground return lead.

PROBLEMS

4.1.   How does the Q of the resonant circuit of a 10Meg/10pF high impedance passive probe with a 6 inch ground lead change when measuring a 50 ohm source if a 100 ohm resistor is added in series with the probe tip? Neglect loading effects of the probe cable shield and assume 20 nh/inch for lead induc-tance. What is the resonant frequency?

Answer: The Q of a tuned circuit is given by $(1/R) \cdot (L/C)^{1/2}$ at resonance. For the 50 ohm circuit alone the Q is about 2.2. With the added 100 ohm tip resistor the Q is lowered to about 0.73. The resonant frequency is about 145 MHz.

4.2   How are the Q and resonant frequency affected in Example 4.1 if a 10Meg/2pF high impedance active probe is used instead?

Answer: The Q for measuring the 50 ohm source is about 4.9 and adding the 100 ohm series tip resistor reduces this to about 1.6. The resonant frequency is raised to about 325 MHz. Note that although the resonant frequency is raised, the Q is signifi-cantly increased by the use of the active probe. Potentially this could lead to increased error at the higher frequencies.

5

High Impedance Passive Probe Compensation

INTRODUCTION

Before leaving the topic of oscilloscope probes and the errors that can occur with their use, probe compensation of high impedance passive probes must be covered. Probe compensation for these probes is familiar to many engineers, yet, it is this familiarity that often causes engineers to overlook this important procedure. Probe compensation is the process of adjusting a probe to deliver its stated accuracy when terminated with the input impedance of the particular scope being used. Most commonly, probe compensation is performed on 10X high impedance passive probes. Passive low impedance and active high impedance probes rarely have user adjustable compensation and this is one of their advantages, for probes which require compensation often are used without proper compensation adjustment. As the reader will see, amplitude errors approaching 100 percent at frequencies as low as 100 kHz and significant waveform errors at higher frequencies can occur.

In this chapter, the reader will learn exactly what probe compensation is and how it is accomplished. The types and magnitudes of errors that can result from improper compensation will be covered.

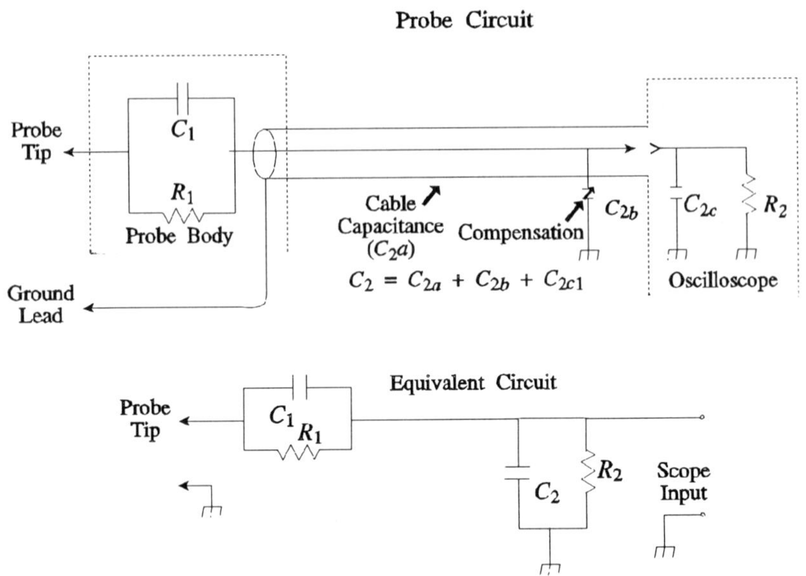

Figure 5.1 Probe equivalent circuit for compensation.

WHAT IS PROBE COMPENSATION?

Common 10X passive high impedance probes utilize a Resistor-Capacitor, RC, voltage divider to obtain the 10:1 voltage division. Compensation is simply the process of adjusting the RC divider so that the probe maintains its 10:1 attenuation ratio over the probe's stated bandwidth. Probe compensation is very sensitive to the input impedance of the scope being used and must be readjusted each time the probe is used with a different scope. It is best to get into the habit of compensating a probe each time it is used. Sometimes even switching two probes between two vertical inputs of the same scope may require them to be recompensated. This is especially true for differential measurements using two probes. This will be discussed at length in the next chapter.

Probe compensation can be understood by reference to the simplified diagram of a probe measurement in Figure 5.1. As mentioned in Chapter 3, the probe tip contains an RC network that, in combination with the scope input impedance and the cable capacitance, forms a 10:1 voltage divider. All 10X passive high impedance probes have the RC network in the probe body and this is represented by R_1 and C_1 in Figure 5.1. The network R_2–C_2 is the parallel combination of the scope input resistance, R_2, the scope input capacitance, and the probe cable capacitance. In addition, either C_1 or C_2 may contain an adjustable capacitor. This adjustable capacitance is the compensation adjustment. In Figure 5.1, V_o is the voltage delivered to the scope input terminals in response to the excitation from the signal source.

The equivalent circuit of the probe in Figure 5.1 is an RC voltage divider. The divider ratio of an RC voltage divider was given in Chapter 2 as:

$$V_o/V_{gen} = \left[\frac{R_2}{R_2 + R_1 \left[\dfrac{1 + s \cdot C_2 \cdot R_2}{1 + s \cdot C_1 \cdot R_1} \right]} \right] \tag{2.24}$$

where $\qquad s = j \cdot 2 \cdot \pi \cdot f.$

The transfer function of the circuit of Figure 5.1 has the same form and is given by Equation 5.1 as:

$$\frac{V_o\,(f)}{V_{gen}\,(f)} = \left[\frac{Z_2}{Z_2 + Z_1} \right] = \left[\frac{R_2}{R_2 + R_1 \cdot \left[\dfrac{1 + s \cdot C_2 \cdot R_2}{1 + s \cdot C_1 \cdot R_1} \right]} \right] \tag{5.1}$$

where: $s = j \cdot 2 \cdot \pi \cdot f,$

$\qquad V_o$ = voltage delivered to scope input terminals,

$\qquad V_{gen}$ = voltage applied between the probe tip and the probe ground connection,

$Z_2 = R_2 \parallel C_2,$

$Z_1 = R_1 \parallel C_1,$

where $\parallel$ means "in parallel with".

At low frequencies, $s < 1/(R_2 \cdot C_2)$, Equation 5.1 reduces to the resistive divider:

$$V_o/V_{gen} = \frac{R_2}{R_2 + R_1} \tag{5.2}$$

and at high frequencies, $s > 1/(R_2 \cdot C_2)$, Equation 5.1 reduces to the capacitive divider:

$$V_o/V_{gen} = \frac{C_1}{C_1 + C_2}. \tag{5.3}$$

It is evident from Equation 5.1 that if the two time constants $R_1 \cdot C_1$ and $R_2 \cdot C_2$ are equal, then frequency dependence disappears from the equation and the probe has a flat frequency response. This condition represents proper compensation of the probe. If the compensation is not properly adjusted and the time constants are not equal, the gain of the scope plus probe is affected at any frequency between 1 kHz and the probe bandwidth.

COMPENSATION AND MEASUREMENT FREQUENCY RESPONSE

The resistances in Figure 5.1 and Equation 5.1 are measured in megohms and the capacitances in tens of pF. This leads to time constants in Equation 5.1 on the order of fractions of milliseconds and corner frequencies of a few kHz. Thus, if a probe is not properly compensated, the probe's attenuation deviates from the intended 10:1 ratio at frequencies starting in the audio band around a few kHz. Measurement errors result throughout the probe's bandwidth.

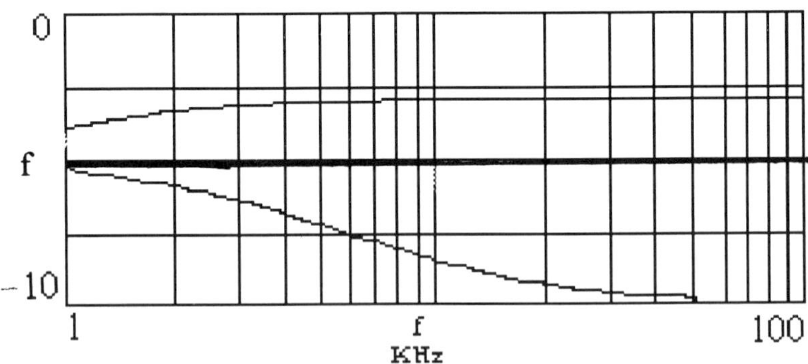

Figure 5.2 Probe frequency response

Figure 5.2 shows the effect of probe compensation on probe frequency response for three cases of probe compensation.

The frequency response of a properly compensated 10X passive high impedance probe, middle trace, an under compensated probe, lower trace, and an over compensated probe, upper trace is shown. The vertical scale is the ratio, in dB from 0 to −40, of the signal delivered to the scope divided by the signal appearing between the probe tip and the probe ground connection as shown in Figure 5.1. The frequency range is plotted on a log scale from 1 kHz to 100 kHz. This frequency range is *very* low compared to frequencies found in many modern circuits. Since a 10X probe is under consideration, a value of −20 dB is the desired probe gain.

The middle trace lies along the −20 dB line which appears slightly darkened. For this case, (properly compensated probe), the probe tip resistor, R_1, is 9 megohms, the probe tip capacitor, C_1, is 11 pF, the scope input resistance, R_2, is 1 megohm, and the sum of the scope input capacitance and the cable capacitance, C_2, is 100 pF. At these low frequencies, the unterminated probe cable can be considered a lumped capacitance and added to the scope input capacitance. These values are similar to those encountered in many modern 10X high impedance probe designs. C_2 is a strong function of cable length and the chosen value represents a probe with a relatively long cable.

As can be seen in Figure 5.2, the frequency response shown for the properly compensated probe is a flat −20 dB over the displayed frequency range of 1 kHz to 100 kHz. This is a result of the two time constants, $R_1 \cdot C_1$ and $R_2 \cdot C_2$, being equal. For these values of R_1, C_1, R_2, and C_2, the probe will meet its gain specification from DC all the way to its stated bandwidth.

Suppose the probe tip capacitor is eliminated and only 1 pF of stray capacitance across the probe tip resistor, R_1, forms C_1. The frequency response for this situation is shown by the lower trace and the probe is described as undercompensated.

Note that above about 60 kHz the probe gain is reduced to −40 dB, the ratio of C_1 to C_2. At all frequencies above 60 kHz, the probe will exhibit this significant error. Some probe designs with the compensation capacitance adjustment in the probe body can be misadjusted by this amount, because the whole value of C_1 is a variable capacitor with a wide range of adjustment.

The upper trace of Figure 5.2 shows an overcompensated 10X high impedance passive probe's frequency response where the probe tip capacitor, C_1, has been increased from 11 pF to 33 pF. The gain error occurs before 10 kHz is reached and is almost 10 dB. Over compensation can also result if C_2 is too small.

COMPENSATION AND WAVEFORM DISTORTION

In addition to frequency response errors, wave shape distortion can also occur because of improper probe compensation. This type of error generally shows up at higher frequencies and is noticeable on a 10 MHz square wave. Figure 5.3 shows a representation of a 10 MHz square wave as displayed on a scope using a common commercially available 10X high impedance passive probe with a 6 foot cable. The two cases shown are for under and over compensation.

Note that undercompensation reduces the amplitude of the square wave while adding a rising edge spike. Over compensation does the opposite. Only for proper compensation does the waveform look correct and have the proper amplitude. Since a real waveform may

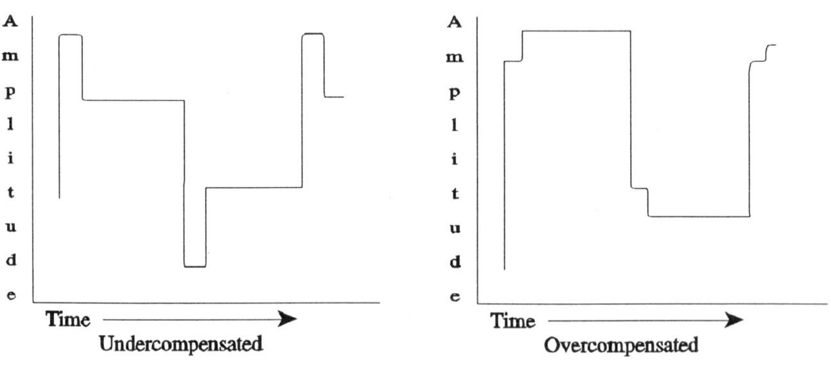

Figure 5.3 Compensation effect on a 10 MHz square wave.

include overshoots, ripples, and possibly an unknown amplitude, it can be difficult to determine when proper compensation has been achieved. The amplitude of the 1 MHz square wave is changed because both the fundamental frequency and all of the harmonics are well above the corner frequencies of the probe circuit as shown in Figure 5.2.

Two probe effects can contribute to the wave shape distortion on the edges. One involves interaction with the probes internal peaking and shaping circuits and the other involves modification of the reflections on the probe cable caused by the different impedance present at one end or the other of the probe cable. While different probes may react differently to improper compensation as it relates to wave shape distortion, it is safe to say that proper compensation is a must for accurate waveform measurements.

COMPENSATION ADJUSTMENT

Although a probe could theoretically be compensated using any signal source within the probe's bandwidth, the use of a 1 kHz square wave has many advantages. These advantages stem from the fact that the corner frequencies of the probe frequency response arising from compensation effects are near 1 kHz. Thus a 1 kHz square wave will

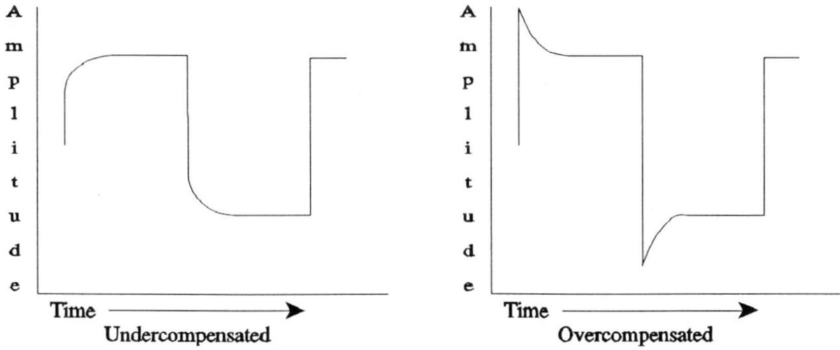

Figure 5.4 Compensation with a 1 kHz squarewave.

have its fundamental frequency component affected differently than the harmonics if the probe is improperly compensated.

The square wave's fundamental is contained in its flat top and bottom, whereas the harmonics are in the rising and falling edges. Thus, if the high frequency response of the probe changes, it will only affect the edges of the square wave and not the flat top and bottom. This makes the difference between the probe's high and low frequency gain easy to see on a low frequency square wave.

For proper compensation, the probe frequency response is flat and the square wave fundamental and all its harmonics are displayed with their proper amplitudes, and thus the wave looks square. An under-compensated probe has poor high frequency response above 1 kHz, see Figure 5.2, and the square wave's harmonics are reduced in amplitude. This effect gives the square wave an undershoot or rounded appearance. The opposite holds for an overcompensated probe, where the square wave harmonics are increased in amplitude relative to the fundamental component, causing an overshoot to appear at each edge of the square wave. The amount of undershoot or overshoot in a 1 kHz square wave represents the amount by which the high frequency gain of the probe is reduced or increased, respectively.

Other signal sources could be used for compensating probes, but none are as easy to use as the 1 kHz square wave. As discussed above, if a 10 MHz square wave were to be used, one would need to know both the amplitude of the wave and exactly how the edges should look in order to adjust the probe. A 100 kHz sine wave would also work. In this case, it would be only necessary to accurately know the amplitude of the source, and the probe would be adjusted to show that amplitude on the scope screen. By comparison, when using a 1 kHz square wave for compensation one needs only to adjust the probe until the signal looks square without needing to know the amplitude of the signal. Some oscilloscopes vary the frequency of the built-in probe calibrator used for compensation with the horizontal sweep frequency. This allows some flexibility to change the frequency around 1 kHz to get a better view of the flatness of the square wave as the exact corner frequencies as shown in Figure 5.2 will vary somewhat with probe design.

COMPENSATION ADJUSTMENT LOCATION

The compensation adjustment for 10X passive high impedance probes is either located in the probe body or in a small box at the scope end of the probe cable. Designs that locate the adjustment at the scope end of the cable have advantages over those that locate the adjustment in the probe body.

EXAMPLE

5.1   For a probe tip capacitance, C_1, of 11 pF, C_2 must equal 99 pF for a 10X probe with a 10 pF input capacitance. If the cable capacitance, C_{2a}, is 75 pF and the compensation capacitor, C_{2b}, is a 1 to 15 pF variable capacitor, then this probe can accommodate scope input capacitances of 9 to 23 pF since C_2 must add up to 99 pF.

When the compensation is located at the scope end of the probe cable it stabilizes the termination impedance at the end of the probe cable. As mentioned in Chapter 3, the probe cable cannot be terminated in its characteristic impedance. It would not be a high impedance probe if it was! Thus the probe design must somehow control the reflections that occur on the cable at higher frequencies, above a few tens of MHz, to insure the probe meets its specifications on impulse response. When the compensation capacitor is located at the scope end of the cable it is adjusted so that the scope input capacitance plus the compensator capacitor always add up to the same total capacitance. This allows the probe designer to plan for a specific load at the end of the cable in the design. If the probe compensation is located in the probe body, the cable termination at the scope will vary from scope to scope depending on the scope input capacitance.

Scope input capacitance can vary from 7 to 30 pF among scopes. Many probes cannot compensate for such a wide range of input capacitance, but a two to one range, say from 12 to 24 pF is not uncommon. A variation from 12 to 24 pF is a substantial change in cable termination that probes with the compensation adjustment at the scope end of the cable do not have to deal with.

Another problem with the compensation adjustment located in the probe body is that frequently all of the capacitance used in parallel with the probe tip resistance is in a single variable capacitance, C_1 as in Figure 5.1. This capacitance can be adjusted over a wide range and thus lead to large errors in probe compensation. Errors in high frequency gain exceeding 100 percent are possible.

Newer probe designs that include the compensation adjustment at the scope end of the probe cable cannot vary the compensation adjustment this much. This is because the compensation capacitor is small compared to the total capacitance of the cable, scope input, and compensation capacitor, on the order of 20 percent of the total for many designs. In fact, the value of the compensation capacitor need only be equal to the range of adjustment of the anticipated scope input capacitance to make the cable plus scope input capacitance plus compensation capacitance always add up to the desired constant value. Errors tend to be limited to around 50 percent or less for this type of design.

Some probe designs have two compensation adjustments. The adjustment discussed so far is called the "low frequency" adjustment and typically is located in the probe body. An additional "high frequency" adjustment is located at the scope end of the cable. This adjustment allows the user to adjust the high frequency impulse response of the probe and is made on a 1 MHz square wave. Potentially, this type of probe could provide more accuracy at the expense of double the effort to compensate the probe.

WHEN TO COMPENSATE

As mentioned earlier, probe compensation should be done every time a probe is connected to a scope. Although this is especially true if the probe is used with different scopes, compensation should be checked even if the probe is always used on the same scope. There can be small differences in the input capacitance of two vertical inputs on the same scope, enough to cause small errors. This effect is especially noticeable when making differential measurements using two probes. This topic will be discussed at length in the next chapter.

SUMMARY

The familiarity of 10X high impedance passive scope probes tends to lead to a complacency on the part of users regarding probe compensation. Points that should be remembered concerning probe compensation are:

1. All frequencies above 10 kHz are affected.
2. Amplitude errors can exceed 100 percent.
3. Wave shape errors also occur, even at high frequencies.
4. If the probe has a single compensation adjustment, generally performance is improved if it is located at the scope end of the probe cable.
5. Check probe compensation at the start of each use.
6. Low impedance passive probes and high impedance active probes usually do not require compensation, a definite advantage.

LABORATORY DEMONSTRATIONS

EXPERIMENTS

Equipment needed:

- 100 MHz or higher bandwidth scope with 1 megohm input impedance and a built in square wave generator for probe compensation,
- Square wave generator capable of at least 5 or 10 MHz such as described in Chapter 3, Figure 3.6, and
- Two 10X passive high impedance probes, one with the compensation at the scope end of the probe cable and the other with the probe compensation in the probe body.

Experiment 5.1: Probe Compensation Effects

1. Connect the probe with the compensation at the scope end of the probe cable to the scope calibrator output and adjust the

horizontal sweep to about 1 ms/division. Set the vertical sensitivity so that the waveform is about one half of the screen height and the triggering for a stable display.

2. Adjust the probe compensation capacitance to produce the maximum overshoot and maximum undershoot.

3. With the compensation adjustment at maximum overshoot, adjust the horizontal sweep speed a few notches slower and faster than 1 ms/division and note the different appearance of the waveform.

4. Connect the probe to the high frequency square wave generator set for about 5 to 10 MHz. Vary the compensation adjustment through its range and note the displayed waveshape.

5. Compensate the probe using the 1 kHz signal from the scope. Note the effect on the displayed square wave when the probe ground lead is removed.

6. Repeat steps 1-4 using the probe with the compensation adjustment in the probe body.

Questions:

5.1.1 What is the maximum percentage of over/undershoot for each of the two probes?

Answer: While the actual maximum amount of undershoot or overshoot is dependant on the probe design, typically one might expect 20 percent to 40 percent for a probe with the compensation at the scope end of the probe cable. For a probe with the compensation in the probe body, the maximum overshoot could exceed 100 percent.

5.1.2 What effect does varying the horizontal sweep speed from .1 to 2 ms/division have on the displayed waveform?

Answer: As the sweep is speeded up to a fraction of a millisecond per division, the exponential decay from an overshoot or undershoot to the flat portion of the top or bottom of the square wave lasts for a greater fraction of the flat portion. Adjustment of the probe compensation is easiest when the decay lasts for about one half of the flat portion of the square wave and only one or two periods of the square wave are displayed on the scope screen. Under these conditions, a slight error in compensation will manifest itself as a slight slope to the top or bottom of the square wave.

5.1.3   What effect does the probe ground lead have on the displayed waveform at 1 kHz?

Answer: The wave shape of the displayed signal will not change. There may possibly be a slight fuzziness to the trace caused by the pickup of signals with a much higher frequency such as a nearby radio station or noise generated by digital logic or a switching power supply. See Example 5.1 for more detail on this effect.

5.1.4   For each of the two probes, how much does the amplitude of the 5 MHz square wave, neglecting for the moment any overshoot on the edges, vary with compensation adjustment?

Answer: The amplitude of the flat portions of a 5 MHz square wave will vary by nearly the same factor as the overshoot or undershoot varies on the 1 kHz square wave used for compensation. See example 5.2 for more detail.

5.1.5   With the 5 MHz square wave, can you tell when the probe is properly compensated?

Answer: It is very difficult. If the amplitude of the 5 MHz square wave and the exact wave shape are known, including any irregularities such as leading edge overshoot, the probe can be compensated by adjusting the compensation to yield the known waveform on the screen. It is much easier to use a 1 kHz square wave.

PROBLEMS AND DISCUSSION

5.1   Many scopes do not even have a place to connect the probe ground lead during compensation. Generally, very little difference will be noted in the waveform if the ground lead is disconnected. After spending all of Chapter 4 discussing the importance of the ground lead, why does it not affect the displayed waveform during probe compensation?

Answer: Ground lead effects are important mostly at frequencies much higher than 1 kHz. Overshoot and ringing due to ground lead effects generally are on a nanosecond to tens of nanoseconds scale affecting the probe response mostly above 20 MHz. If the scope 1 kHz square wave has fast enough edges, these effects will be present during compensation even though the calibration source and scope amplifier are referenced to the same ground – the scope ground. However, with the sweep set for 1 ms/division, they will not be visible.

5.2   If a probe is not properly compensated so that a 1 kHz square wave contains a 20 percent overshoot, how many dB is the 100 kHz gain of the probe changed by? Is it higher or lower than it should be?

Answer: The 100 kHz gain will be increased by 20 percent so the amount of increase in dB is 20log(1.2) or about 1.6 dB.

6

Differential Measurements

INTRODUCTION

What Is A Differential Measurement?

Any voltage measurement must by definition require two nodes for the measurement. For some measurements, one of the nodes is ground while for others neither is ground. Occasionally, both measurement nodes are *ground*. But in all cases what is desired is the voltage difference between the two measurement nodes.

Up to this point all of the measurements discussed have been *unbalanced*. That is, one side of the measurement was ground, usually the scope probe ground clip. Chapter 4 pointed out some of the errors that can occur because of currents flowing between the EUT ground and the scope ground over the probe ground lead. In a differential measurement, both measurement leads from the scope probe or probes have the same impedance to the scope ground and it does not matter which node, if any, of the EUT is ground. Ideally the scope responds only to the difference in voltage between the two nodes. Any potential difference between the EUT ground and the scope ground is rejected as a common mode signal.

Why Use Differential Measurement Techniques?

Differential techniques for measuring voltages in circuits offer several advantages that result in increased accuracy of measurements over the unbalanced scope probe (referenced to ground) measurements discussed so far. These techniques remove many of the limitations of unbalanced measurements, if done properly. This chapter will discuss these advantages as well as other aspects of differential measurements including options available for making differential measurements and the proper techniques to use. The use of proper techniques for making differential measurements is critical and will be discussed in detail.

The null experiment cannot be dispensed with when making differential measurements. In fact, validation of measurement results by an appropriate null experiment is even more critical for differential measurements than it is for unbalanced measurements. Later in this chapter it will be shown how it is possible to get common mode *gain* rather than common mode rejection in some seemingly simple differential measurements. Appropriate null experiments to prevent effects, such as lack of common mode rejection, from introducing error into measurements will be discussed.

Construction details will be discussed for a high performance balanced coaxial probe than can be built in about 30 minutes with excellent performance on many common types of differential measurements. Measured data taken with this probe will be compared to that taken with conventional unbalanced scope probes.

ADVANTAGES OF DIFFERENTIAL
MEASUREMENTS

The most important advantages of differential measurements include:

1.   Minimization of ground loading error,
2.   Rejection of ground lead (if used) induced voltage, and
3.   Common mode rejection of other sources of interference.

As mentioned in Chapter 4, ground loading occurs when a lead, such as a probe ground lead, is connected to the circuit ground of the

Equipment Under Test, EUT. High frequency potential relative to the equipment chassis or ground is almost inevitably present on the circuit ground of most modern electronic equipment. It will drive a noise current into any lead connected to the circuit ground. One effect of that current is to change the distribution of high frequency currents in the equipment circuit ground which can affect equipment operation. As will be shown in Chapter 9, a lead connected to an EUT circuit ground can also conduct interference into the equipment.

For many differential measurements, the probe ground lead or leads are not used at all. The lack of a hard connection to the EUT circuit ground minimizes the effects of ground loading.

If a probe ground lead is used, the voltage induced in it by noise currents from circuit ground can, with proper design, be made to appear as a common mode voltage to be rejected in a differential measurement. The criteria for deciding when to use a ground lead will be discussed later in this chapter.

In addition to rejecting ground lead voltages, other sources of interference tend to be common mode on the two nodes between which the measurement is being made. This rejection of external interference is especially important when making single shot measurements of transient signals with a fast digital storage scope. An unbalanced measurement of this type is doomed to failure by any number of noise generators such as transients due to switches on the AC power line, Electrical Fast Transient, and Electrostatic Discharge events unlikely to be noticed by the person making the measurement.

With all these advantages, differential voltage measurements become the preferred method for many measurement needs. *But,* keep in mind that it is very easy to lose all the advantages of a differential measurement if the measurement is not done properly.

AVAILABLE OPTIONS FOR MAKING DIFFERENTIAL MEASUREMENTS

FET Differential Probes

There are several ways of making differential voltage measurements. One popular way is to use a commercially manufactured active differential probe. Probes of this type were mentioned in Chapter 3. There are only a few units commercially available at the time of this writing suitable for measurements above a few tens of MHz. A popular model that has been around for many years uses FET transistors and other circuitry in the differential probe head. The probe head is connected by a cable carrying power to the active circuitry in the probe head and the signal to a separate amplifier whose output is connected to a conventional oscilloscope. The performance of this unit is reasonable, giving 40 dB of common mode rejection at 100 MHz. But there are disadvantages to using an active differential probe such as the one described above. One of the most important disadvantages is the lack of robustness in some probe designs.

The **Mean Time Before Failure**, MTBF, of a probe is dependent upon both its design and how it is used. Contributing factors to a short probe MTBF are: overload damage or otherwise exceeding the electrical ratings of the probe, running over the probe with the wheel of a scope cart, and other physical stress to parts of the probe such as the probe tip. In my experience, probe MTBFs range from a few days to many years. In general, it is desirable to use a robust probe that does not require special care if possible.

Some active FET differential probes are very fragile and thus have a short MTBF. Given the care that many probes get in a typical research and development laboratory, some differential probes would have a MTBF of about 30 minutes.

Actually a fragile instrument of any type has the disadvantage of not being available at all times for use. It will be in the shop for repair instead of on the lab bench.

Some new FET differential probe designs incorporate easily replaceable parts for those that might be easily broken and incorporate input protection as well, reducing the need to send the probe to the manufacturer for service. The usable frequency range is still limited to about 100 to 200 MHz though and input protection adds input capacitance and its resultant circuit loading. A popular model of modern design has an input capacitance of 7 pF, almost as much as a passive 10X high impedance probe.

Even with a robust differential probe, another problem is cost. A cost of a few thousand dollars is typical. Probes costing this much will not be as commonly available in a laboratory as standard probe types costing only a hundred dollars or so.

Differential FET probes do have several advantages that make their cost justifiable in some measurement situations. For one, some of these probes have a relatively high input impedance, on the order of 1 megohm in parallel with as little as a couple of pF of capacitance (note that 2 pF has a reactance of about 800 ohms at 100 MHz). Another advantage is good common mode rejection, on the order of 40 dB, over the specified frequency range which can reach as high as 100 MHz.

There are many jobs for which an FET differential probe is necessary. Some of these applications require a probe to have a high input impedance to moderately high frequencies. Other applications need the good common mode rejection offered by the FET differential probe. However, for many applications there are other alternatives that will get the job done at a much lower cost.

Two Hi-Z 10X Passive Probes Using A-B

One of the alternatives to using an FET differential probe is to use two standard 10X high impedance passive probes, one on each channel of a dual channel scope, with the scope set to subtract the two channels (A-B setting). This arrangement offers the convenience of commonly available, inexpensive probes. For best results, the two probes used should be matched. For many reasons to be discussed shortly, this arrangement should not be used at frequencies above a few MHz. Now, let's examine the requirements for accurate differential measurements with two separate probes.

First, the lengths of the two cables must be equal. If there is a difference in length, the signal delay from the probe tip to the scope amplifier input will be different for the two probes. This will result in poor common mode rejection for fast rise times. Consider a square wave that is fed common mode to both probes with a two nanosecond (ns) risetime and one probe has a cable one meter longer than the other. This would represent almost a six nanosecond time delay difference between the probes and would cause degradation of common mode rejection in this case. A rising edge of the square wave would be completed as seen by the scope on the short probe before the longer probe's scope channel would see the start of the rising edge. Thus for every transition of the common mode signal, the scope will display a pulse of about 6 ns duration.

Since the scope is subtracting two nearly equal numbers, if it is to have common mode rejection, probe compensation is very critical, much more so than in single ended measurements. The compensation process must include extra steps to balance the two probes as follows:

1. Compensate each probe separately by the standard procedure using a 1 kHz square wave. Then set up the scope in the A-B mode using both probes.
2. Connect both probes to the 1 kHz square wave source and adjust either one (not both) slightly for the best null. The purpose of this step ensures that they are both compensated exactly the same by adjusting one to match the other. It does not matter which probe is adjusted since they were nearly the same after step 1.

Another important check that must be done is a null experiment. For the case of using two probes to make a differential measurement, connect the two probe tips together and then connect this node to the

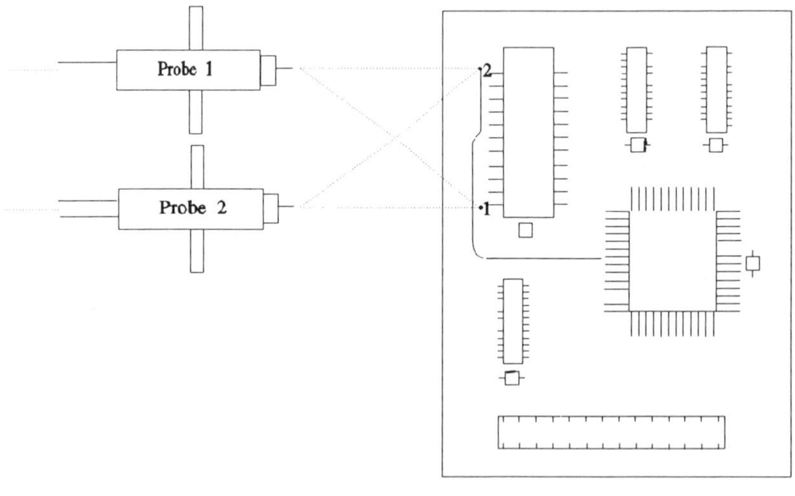

Figure 6.1 Checking common mode response.

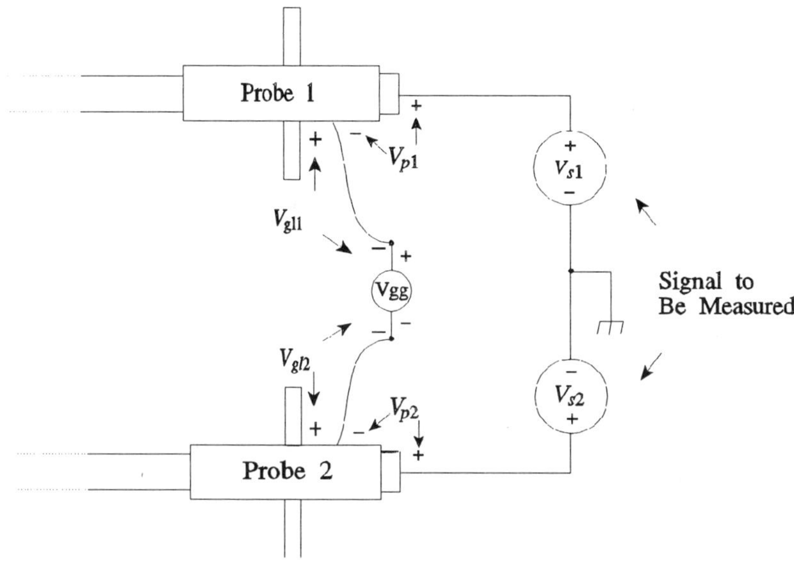

Figure 6.2 Differential ground lead error.

two points to be measured, one at a time. The maximum displayed signal at the two points to be measured is the common mode response of the probes and represents the margin of error of the measurement. The process is illustrated in Figure 6.1.

In Figure 6.1, it is desired to measure the high frequency voltage on a Printed Wiring Board, PWB, between nodes 1 and 2. Note that node 1 is a "ground" and node 2 is a signal. The probe tips are first shorted to each other and then touched first to node 1 and then to node 2. The peak value of the displayed waveform, ideally zero, for each node is noted and the highest one becomes the margin of error for the measurement. The allowable margin of error depends on the measurement to be made. For noise measurements, a 10 percent error is usually acceptable.

The treatment of the probe ground leads will have a direct affect on the accuracy of the measurement. The two probe ground leads should be connected together and either floated or connected to circuit ground at one point. The decision to float the ground leads or connect them to ground will be discussed shortly. If the ground leads are used, it may seem logical to connect each one to ground on the circuit being tested physically close to its probe tip. However, any potential that exists between the two probe ground lead clips will show up directly as an input to the probes and appear as part of the signal being measured. Figure 6.2 illustrates the problem.

The total voltage, V_t, around the measurement loop is given by:

$$V_t = V_{s1} - V_{s2} - V_{gl1} + V_{gl2} - V_{gg} - V_{p1} + V_{p2} \qquad (6.1)$$

where: V_{s1} and V_{s2} are the signal voltages to be measured,
V_{gl1} and V_{gl2} are the voltages across the two ground leads,
V_{gg} is the voltage between the ends of the ground leads, and
V_{p1} and V_{p2} are the voltages across the probe inputs.

Since the total voltage around the measurement loop must equal zero, equation 6.1 may be solved for the probe input voltages as follows:

$$V_{p1} - V_{p2} = V_{s1} - V_{s2} - V_{gl1} + V_{gl2} - V_{gg} \qquad (6.2)$$

Equation 6.2 shows that the scope will display not only the difference between the two signal voltages, but the difference between the ground lead voltage drops and the voltage between the ends of the ground leads as well. The only way to ensure that the voltage between the ends of the ground leads is zero is to connect the ground leads of the two probes together.

Whether the ground leads, connected together, should be left floating or connected to ground in the EUT is determined by the amplitude of the common mode voltage delivered to the scope by the probes. When making a differential measurement using two probes, first check that the displayed signal for each channel on the scope is on screen. If either channel is driving its trace off the scope screen, the vertical amplifier for that channel may be saturating. Vertical amplifier saturation will cause error to occur when the A-B subtraction is done and will result in the scope displaying an incorrect differential signal. If the common mode signal is driving either trace off the scope screen then the scope vertical sensitivity must be reduced. If the differential signal is small, reducing the vertical sensitivity may preclude making the measurement. In this case, the probe ground leads should be connected to one point on the EUT ground to reduce the common mode signal amplitude. Of course, this comes with the penalty of introducing ground loading and the difference in the ground lead induced voltages as well into the measurement, but sometimes it is the only way a particular measurement can be made.

Reduction of common mode signal is the only reason for connecting the probe ground lead(s) to the EUT ground when making a differential measurement. This is true whether the measurement is an A-B measurement using two probes or one using a true differential probe. If possible, it is preferable to leave the ground lead(s) floating (connected together if there are two).

There is another source of error in differential measurements using two probes with separate ground leads. In Equation 6.2, it was shown that the differential signal displayed by the scope included the difference in the voltages across the two ground leads. There are two ways in which the individual ground lead voltages can add a significant error signal into the differential measurement.

One way involves direct connection of the EUT ground to the probe ground leads. If the ground leads are attached to the EUT ground to reduce common mode signal, then noise currents will be driven from the EUT ground into each ground lead. The driving voltage is the same for both leads since they are connected together,

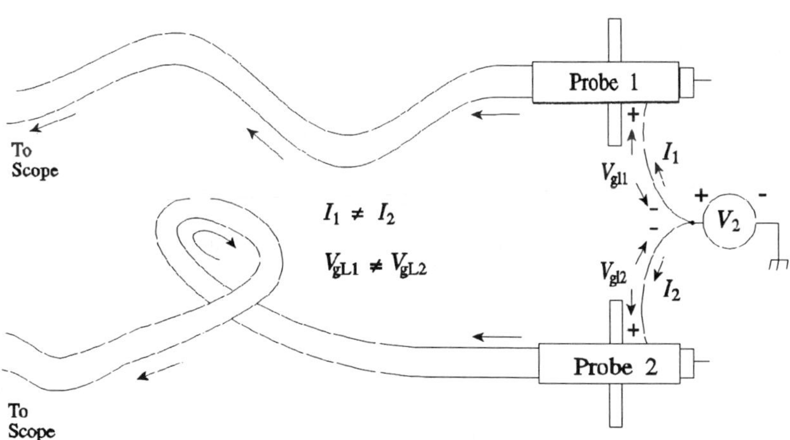

Figure 6.3 Ground lead induced error (ground leads grounded to EUT).

but the impedance this driving voltage sees is different for each lead. Figure 6.3 depicts the situation.

Many *many* factors contribute to the different impedances seen looking into the probe ground leads. Most are related to the fact that the two probe cable shields are not positioned precisely the same. For instance, one may have an extra loop or bend in it or one may be closer to some grounded metal increasing its capacitance to ground compared to the other cable shield. In addition, even if the ground lead currents were the same, their inductances would be slightly different due to differing ground lead shapes and positions. The net result is that the two ground lead induced voltages are not the same and this difference signal adds directly to the differential signal being measured.

The other way probe ground lead induced voltages can add error to a differential measurement involves circulating currents in the large loop composed of the probe cable shields, the scope chassis, and the probe ground wires. These currents can be induced by high frequency magnetic fields passing through this loop even if the probe ground leads are floating. Such circulating currents contribute error to the measurement by generating probe ground lead voltages that

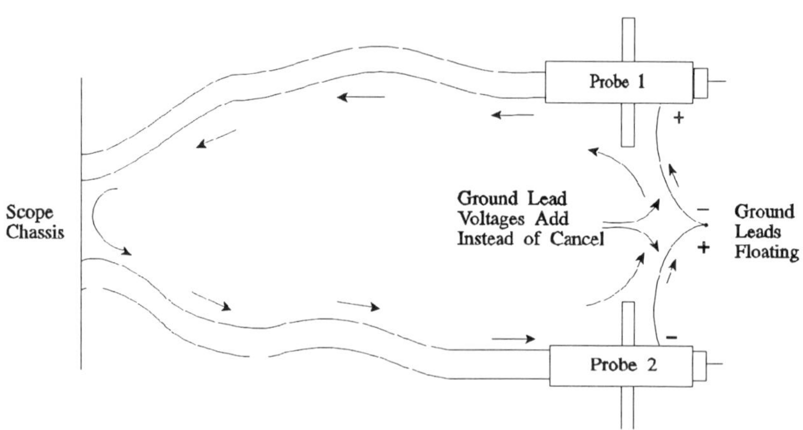

Figure 6.4 Ground lead induced error
(ground leads floating).

add to the differential signal instead of cancelling each other. Figure 6.4 shows how this can happen.

As can be seen in Figure 6.4, the voltages induced on the ground leads by the circulating current add to the measured voltage instead of cancelling. This can be seen with reference to Equation 6.2 repeated below.

$$V_{p1} - V_{p2} = V_{s1} - V_{s2} - V_{gl1} + V_{gl2} - V_{gg} \qquad (6.2)$$

The ground lead voltages are defined in Equation 6.2 as $+$ to $-$ from the tip of the probe ground leads to their respective probe bodies. But, the circulating current induces a voltage on the second ground lead that is negative by the above definition, so that the two ground lead voltages add. For this case, one could rewrite Equation 6.2 for the worst case as:

$$V_{p1} - V_{p2} = V_{s1} - V_{s2} + |V_{gl1}| + |V_{gl2}| \qquad (6.3)$$

where V_{gg} is zero since the ground leads are connected and the assumption is made that since the ground leads are short there are no wavelength associated phase shifts in the currents between the ground leads.

The magnitude of the circulating current and therefore the ground lead voltages can be reduced by reducing the area of the loop to receive the magnetic field. This can be effectively accomplished by fastening the two probe cables together with cable ties every 6 inches or so. The probe cables should *always* be tied together this way when making a differential measurement using two separate probes.

The limitations posed by the induced ground lead voltages cause the **C**ommon **M**ode **R**ejection **R**atio (CMRR) of a two probe A-B differential measurement to rarely exceed 10 dB at frequencies above a few tens of MHz, especially if the ground leads are connected to the equipment ground. In fact, as will be shown in the experiments at the end of this chapter, the CMRR can approach zero and even become common mode *gain* in extreme cases. Remember, even if the ground leads are floated, this type of differential measurement will be sensitive to local fields generating circulating currents in the loop formed by the probes. For these reasons differential measurements involving frequencies above a few MHz should not be made using two separate probes.

The Balanced Coaxial Probe

There is a simple probe that can be easily constructed that performs well with good common mode rejection into the hundreds of MHz and is not sensitive to ground currents. It is called a balanced coaxial probe and is shown in Figure 6.5.

This probe is a high performance low impedance, 1000 ohms center-tapped to ground, differential probe. Although not yet available commercially, it is easily built from commonly available parts. It is constructed from a pair of 50 ohm coaxial cables with male BNC connectors on one end in the following way:

1.   Strip the outer insulation from both cables starting about six inches from the BNC connector.
2.   Solder the exposed shields together every 3 to 6 inches.
3.   Fasten a resistor, usually 450 ohms, to the center conductor of each coax and a single ground lead to the shields where they end. The individual cable shields should extend all the way to the end of the cable where the resistor is attached, but left open

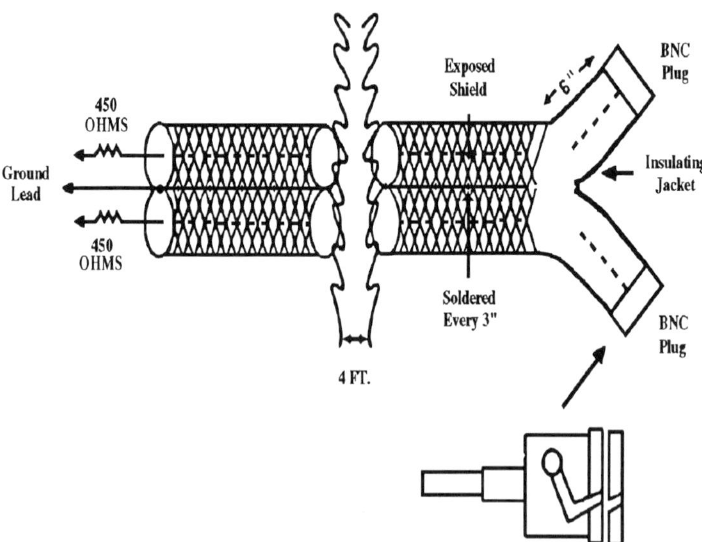

Figure 6.5 Balanced coaxial probe.

circuited. This will minimize the effect of hand capacitance on the probe operation.

4.   Cover the exposed shields with heat shrink tubing to prevent shorts to circuitry in the equipment to be measured.

The reader may remember from an earlier chapter that each half of this probe is a low impedance passive probe. The 450 ohm resistors form a 10:1 divider with the 50 ohm impedance of the cable, although other values of resistors may be used to achieve different division ratios. Also remember that the input impedance of a "low" impedance probe can be substantially higher than that of a "high" impedance passive probe at frequencies above a few tens of MHz. Therefore the 1000 ohm input impedance is relatively high compared to that offered by using two passive high impedance probes in an A-B measurement as described earlier.

Common mode rejection of ground lead induced voltages and rejection of interference from local fields are enhanced in this probe by several factors. First, there is only one ground lead and therefore its induced voltage *must* be common mode to both channels. Second, by keeping the shields in contact, loop area between the shields is minimized thus reducing any circulating currents induced by local fields, and the two shield currents are made nearly equal. Equal shield currents help improve the CMRR by balancing any contribution to the signal delivered to the scope through the shield transfer impedance[1] of the coaxial cables. And third, the two leads are always located in precisely the same position.

Performance of this probe is very good. Measured results of 25 to 35 dB CMRR at 100 MHz has been achieved without taking unusual steps in the construction of the probe. Moderate care in construction is necessary though. Resistor tolerances of 1 percent or better should

---

1.   The shield transfer impedance of a cable relates the shield current to the open cicuit voltage (per unit length) generated between the center conductor and shield to the shield current. Shield transfer impedance has units of ohms/meter. For more information see *Noise Reduction Techniques in Electronic Systems,* H. W. Ott, Second Edition, John Wiley & Sons, Inc., pp. 55-59.

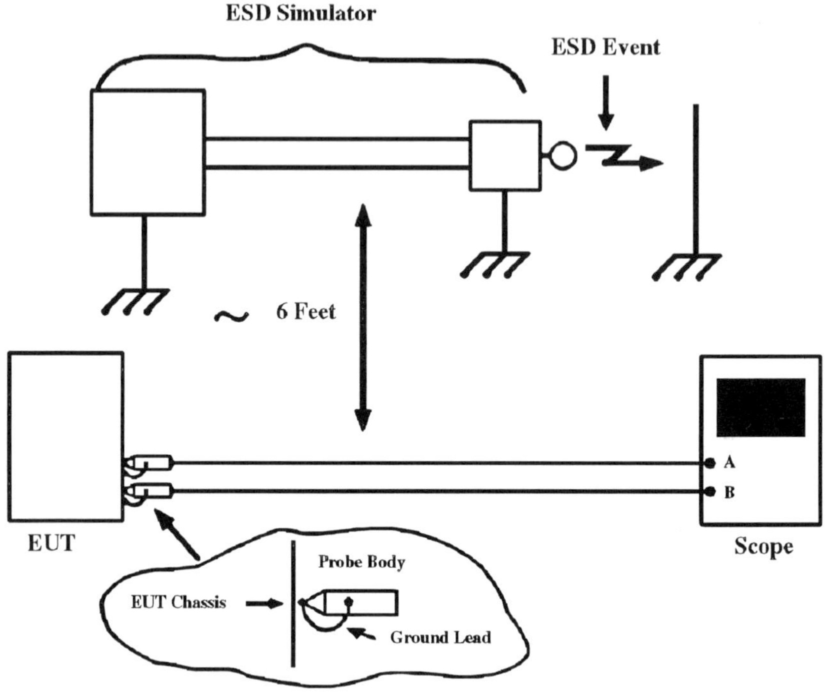

Figure 6.6 Experimental setup for probe test.

be used and symmetry in the construction of the probe should be maintained. Of course, wire wound resistors should be avoided.

ElectroStatic Discharge, ESD, is a potent source of interference to measurements. One example of this probe's performance compared to a pair of 10X passive high impedance scope probes can be seen in the following experiment involving ESD generated noise[2]. Figure 6.6 shows the experimental setup for the two conventional probes.

---

2. Figures 6.6-6.9 reprinted with permission of the EOS/EDS Association, Inc. "Applying High Frequency Techniques to Measuring ESD Phenomena and Its Effects.", D. C. Smith, 1989 EOS/ESD Symposium Proceedings, EOS-11, pp. 114-119.

In this test, the common mode response of two 10X high imped-ance probes is evaluated by shorting each probe using its own ground lead and connecting both shorted probes to the same point on the chassis of a nearby piece of equipment, an EUT. An ESD simulator is used to generate ESD events about 6 feet away from the test setup. The output of each probe is displayed on a separate trace of the scope. Since the probes are shorted, at first glance one would expect that the scope would display no signal at all. But, this is not the case.

Figure 6.7 shows the response of each shorted probe separately as displayed on the channels of a dual channel digital storage scope. Each channel is displayed with a vertical scale of ± 8 Volts and a horizontal scale for time from 0 to 200 ns.

Not only are the probe responses nonzero, reaching peak values of 4 to 8 volts, they are out of phase at some points in time, causing the difference between the channels to exceed the peak value of either one. At 50 ns, the difference between the two channels is almost 11 volts, significantly exceeding the highest peak value of channel A or B effectively producing common-mode gain. Clearly there is nothing common mode about the interference to these two probes. As ex-plained earlier, this is partially due to the differing ground lead currents coming from the chassis they are fastened to and partially from induction from the fields emanating from the ESD event into the loop formed between the two probe cables.

The results for the same experiment using the balanced coaxial probe is shown in Figure 6.8. The only difference in the experiment is the use of the balanced coaxial probe. Its two probe tips, the 450 ohm resistors, are connected together to the probe ground lead and this combination is connected to the chassis of the nearby EUT.

Two characteristics of the response shown in Figure 6.8 imme-diately stand out. First, the two channels are displaying about the same waveform and, second, the two waveforms are in phase. There is a slight difference in the two waveforms, channel B has slightly more fine detail in it. This difference was traced down to the channel B vertical amplifier. In any event, the interference is mostly common mode and can be subtracted out in an A-B measurement.

To underscore the variability of the common mode interference in this test setup. Figure 6.9 shows the same test that resulted in the waveforms of Figure 6.8, except the middle of the cable was moved about 1 foot. Notice how drastically the response of the two channels

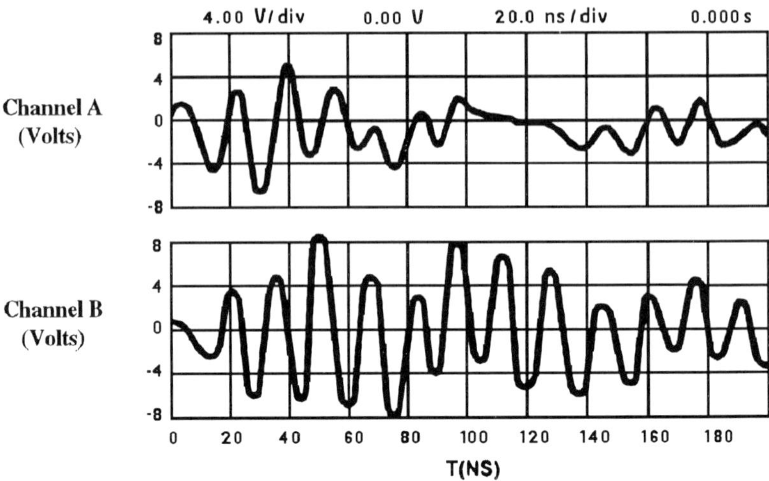

Figure 6.7 Shorted probe response of two adjacent probes.

has changed, but it is still common mode and therefore can be subtracted out from the measurement.

The main limitation on the use of this probe concerns its input impedance. Since it looks like 1000 ohms center-tapped to ground for a 10X version, the balanced coaxial probe should not be used to make measurements on high impedance circuits. This is not as much of a problem as it appears since high impedance circuits are relatively rare in high frequency circuits. Note that 10 pF of capacitance has a reactance of only 160 ohms at 100 MHz. Above several tens of MHz, most circuits have an impedance of a few hundred ohms or less, the main exception are high Q resonant circuits. Tuned resonant circuits should not be measured with this probe.

Most noise measurements on printed wiring boards are on low impedance circuits. For example, if the power to ground impedance on a printed wiring board were comparable to the 1000 ohm input impedance of the balanced coaxial probe, the board would likely not work at all. Similarly, the ground to ground impedance of a PWB should be relatively low as well.

PROBE CONNECTION TECHNIQUES

Connecting scope probes to a printed circuit board to measure a voltage can offer a challenge if accuracy is to be preserved. There are several ways to do this. Each will be discussed in this section, in order of increasing accuracy, ending with an easy way to use the balanced coaxial probe for measurements on a printed wiring board.

For unbalanced probes, finding a convenient connection point for the probe ground lead on a printed wiring board can be a major problem. Often the connection is made at the board edge, or at best some distance from the measurement to be made. The result can be a long ground lead leading to probe resonance and the addition of the board ground noise into the measurement circuit.

One method to shorten the required probe ground lead is to tack solder short leads, about ½ to 1 inch, in pairs at the location of each measurement to be made, one stub on ground and the other on the node to be measured. In addition to shortening the leads, the process of measuring proceeds more quickly and with less chance for error by eliminating the need to constantly refer to drawings as the board measurements are being made. The stubs clearly mark the locations to be measured.

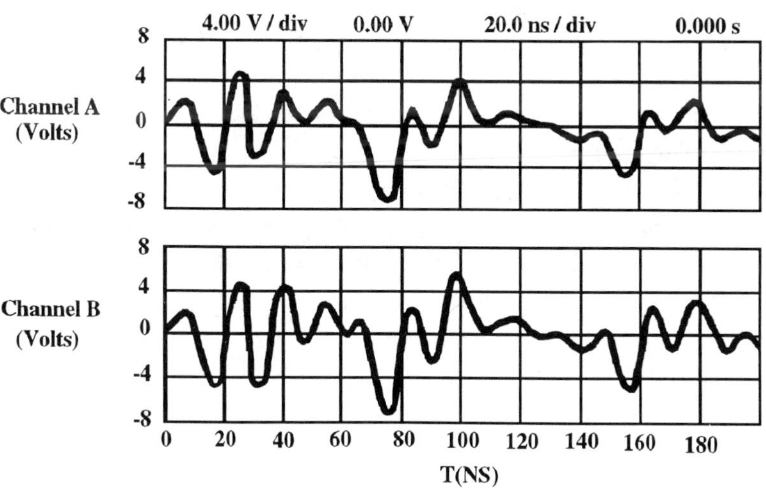

Figure 6.8 Balanced coaxial probe, shorted probe response.

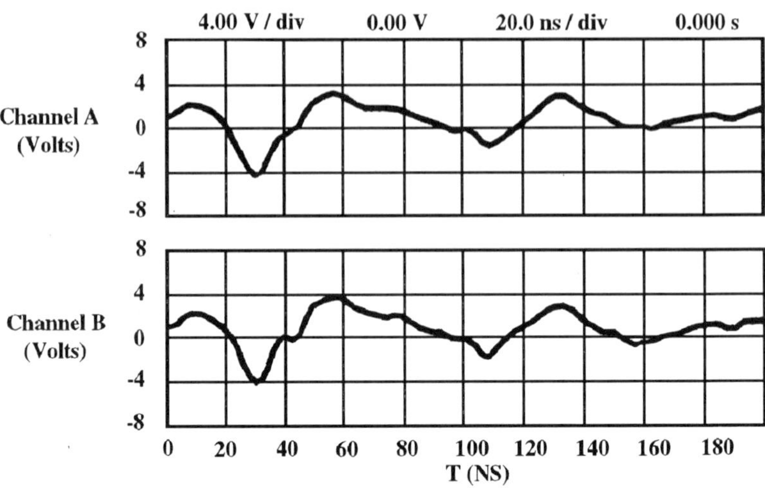

Figure 6.9 Balanced coaxial probe, shorted probe response (middle of cable moved).

If the measurements are to be made with a high impedance passive probe, consider replacing the stubs attached to the signal nodes with 100 ohm carbon composition resistors as discussed in Chapter 4. As discussed earlier under probe resonance, this will generally improve measurement accuracy. As an added benefit, the signal nodes will now be visually distinguished from the ground nodes. This can avert the unexpected surprise of connecting the probe ground lead to the board power grid when making power to ground noise measurements.

One could also use a small coaxial probe socket available from some of the major probe manufacturers. These small sockets mount on the circuit board and have very short ground and signal connections. The probe plugs directly into the socket. The center pin of the probe inserts into a pin receptacle and flexible contacts make contact to the ground ring near the end of the probe for a full 360 degrees. Probe sockets of this type are best used for permanent mounting on a circuit board for frequently made measurements or where accuracy with high impedance passive probes is paramount.

Probably the best way to make signal or noise measurements on printed wiring boards in most cases is with the balanced low impedance coaxial probe. For this probe, replace the stubs soldered to the printed wiring board with 450 ohm resistors for both signal and ground nodes (if part of the measurement). The end of the coaxial cables of the probe are equipped with pin jacks which are plugged onto the resistors to make a measurement. Note that it is not necessary to distinguish between signal and ground nodes for this balanced measurement.

SUMMARY

Differential measurements have significant advantages over unbalanced measurements. Some of the more important points to remember about differential measurements are:

1. The ground lead can (and should if possible) be floated. This reduces ground loading on the EUT and may improve common mode rejection.
2. Common mode overloading of the scope vertical amplifiers must be avoided when using two separate probes or the balanced coaxial probe. Make sure that each of the channels displayed separately remains on the scope screen.
3. Always check the common mode response of a differential measurement by doing a null experiment consisting of connecting the two probe tips together and then connecting this combination to each of the two points to be measured individually. The waveform that results from this experiment is the margin of error of the differential measurement.
4. Use of two separate scope probes should be limited to low frequency measurements, less than a few MHz. At higher frequencies, the common mode rejection falls rapidly to values that cause unacceptable error in the measurement.
5. The balanced coaxial probe works well at low cost for many signal and noise measurements on low impedance circuits. Don't forget the null experiment.
6. Differential FET probes must be used for high impedance circuits. Be sure to check for common mode overloading, the null experiment is a must.

LABORATORY DEMONSTRATIONS

EXPERIMENTS

Equipment needed:

- 100 MHz or better oscilloscope (analog preferred),
- 5-50 MHz square wave oscillator (see Figure 3.6),
- Two high impedance 10X passive probes with 2 meter cables,
- One balanced coaxial probe, and
- Several 1 foot test clips.

Experiment 6.1: Effect of Using Probe Ground Leads on Common Mode Rejection

1. Connect the two 10X passive high impedance probes to the vertical inputs of the scope and set the vertical amplifiers to perform a channel A minus channel B measurement. Be sure to properly compensate both probes and then adjust one of them for a null with both probes connected to the compensation

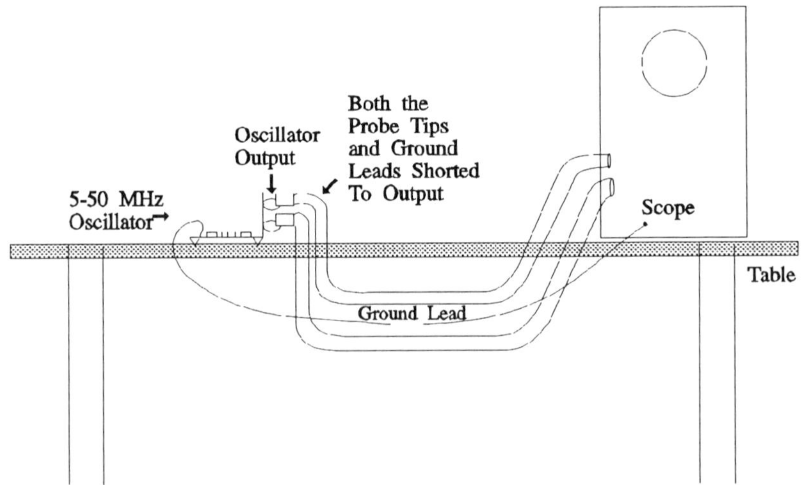

Figure for Example 6.1 Probe and scope connections.

generator as discussed earlier in this chapter. Set the individual vertical channels of the scope to 1 volt per division (don't forget the 10X factor of the probes).

2. Connect the ground of the 5-50 MHz square wave oscillator to the chassis of the scope through about 3 to 5 feet of test lead as shown in Figure for Experiment 6.1 below. Put a 1 inch stub on the output of the oscillator and connect both probe tips and both probe ground leads to this stub. Dress the probe cables near the ground return lead of the oscillator. This connection causes the oscillator to drive a signal current into the probe shields, onto the scope chassis, and back through the test lead to the oscillator "ground." The current is not constant in this path because the dimensions of the loop comprised of the probe leads and the test lead connection from the scope to the oscillator ground are large enough to permit some of the current to be radiated as well as standing waves to be formed on the loop.

3. Turn on the oscillator and vary the frequency between 20 and 50 MHz. It should be possible to move the probe cables and the test lead ground wire so that at some frequency in the 20 to 50 MHz range the A-B display on the scope will read 2 volts peak or more. Set the vertical amplifiers for A+B operation. Is the displayed waveform larger or smaller? Estimate the common mode rejection of this measurement.

Questions:

6.1.1 Why does this happen?

Answer: The oscillator drives both probe ground leads and cable shields with the same voltage, but the impedance looking into each probe ground lead is different depending on the exact position of the probe cable and its proximity to nearby objects. Thus, the current driven into each probe ground lead will be different and thus the $L \cdot dt/dt$ drop across each of the two probe ground leads will be different and the voltages will not cancel in the A-B measurement. In fact, it is possible for the ground lead voltages to be out of phase and add up when the A-B operation is performed in the scope leading to common mode gain.

6.1.2   Most measurements, especially on digital equipment, are not usually of a single frequency oscillator so why is the effect apparent at discrete frequencies in this experiment a problem?

Answer: Most digital equipment contains broadband noise, so if the measurement setup has a peaked response at one frequency, there will be some energy available to excite this measurement response and induce error into the measurement.

6.1.3   What is the common mode rejection in this measurement?

Answer: The common mode rejection, which can be defined as the ratio of the peak-to-peak magnitude of the A-B signal to the peak-to-peak magnitude of the A+B signal can range, at the worst frequencies, from a few dB to a few dB negative (common mode gain). Note that the A-B signal displayed on the scope can have a peak-to-peak magnitude approaching the open circuit voltage of the oscillator.

Experiment 6.2: Balanced Coaxial Probe Performance

1.   Connect a 10X (500 ohm) balanced coaxial probe to the 50 ohm inputs of a scope. If the scope only has 1 meg inputs, use a 50 ohm BNC termination resistor at each scope input.
2.   Connect the oscillator ground to the scope chassis and put a 1 inch stub on the oscillator output as in Experiment 6.1 above. Connect both 450 ohm resistor inputs of the probe and its ground lead to the stub. Dress the probe cable near the ground return lead.
3.   Turn on the oscillator and vary the frequency between 20 and 50 MHz with the scope vertical sensitivity set to 1 volt per division as in Experiment 6.1 above. Vary the positions of the leads and frequency to achieve the maximum A-B signal and note the amplitude of this signal as well as the effect of changing the frequency. Repeat using the A+B setting on the scope vertical amplifier.

Questions:

6.2.1 How do the displayed A-B signal as well as the A and B signals individually compare to Experiment 6.1 as the frequency is varied?

Answer: The amplitude of A-B is much smaller, on the order of tens of millivolts instead of volts. In addition, the amplitude does not peak up substantially at discrete frequencies, but remains nearly constant over frequency. This is due to the resistive input impedance of the balanced coaxial probe. The reactive input impedance of the passive 10X high impedance probes in Experiment 6.1 in combination with the ground lead inductance and other circuit impedances can cause resonance conditions to exist that are minimized with the balanced coaxial probe. This causes the A and B signals to have smaller amplitudes for the same length ground lead than the high impedance passive probes.

6.2.2 How does the A+B signal contrast to Experiment 6.1?

Answer: The A+B signal in this experiment is about 25 dB or more greater than the A-B signal (due to the better common mode rejection of the balanced coaxial probe). Even so, it is only about one half or less of the A+B signal in Experiment 6.1 because of the resistive input impedance of the probe. The resistive input impedance of the balanced coaxial probe is a major advantage over passive high impedance probes in many measurement situations.

PROBLEMS AND DISCUSSION

6.1. Using two separate probes and a digital scope, a differential measurement of a ground to ground noise voltage on a digital circuit is made. Assuming that the common mode rejection is adequate for this measurement (a questionable assumption), data is taken. If each probe has a 4 inch ground lead and the ground leads are connected together and floated, what error is introduced into the measurement by a magnetic field that induces a *di/dt* of 25 ma/ns into the loop formed by the two probe cables and the ground leads?

Answer: Only the current flowing in the ground leads will substantially contribute to measurement error for well designed probes. The voltage induced on the ground leads by the circulating current from the interfering magnetic field will be an error for this measurement. One can estimate that voltage as follows:

$$V = L \cdot di/dt \tag{6.4}$$
$$L \approx 20 \ nh/inch \cdot 8 \ inches \ (of \ ground \ lead) = 160 \ nh$$
$$di/dt = 25 \ ma/ns \ (a \ given)$$
$$therefore:$$
$$V \approx 4 \ volts.$$

This magnitude of voltage can easily be produced by an electrostatic discharge event from across a room and points out a major concern about using digital scopes. That is, even if the common mode response were acceptable in this example, transient common and differential mode signals can be introduced by ESD events and other phenomena not associated with the measurement being made. Analog scopes without storage will not display such transient events but digital scopes will display them with the same clarity as the intended measurement. This subject will be treated in greater detail later in Chapter 9.

6.2 Ground loading can be minimized in a differential measurement by eliminating the ground lead as long as the common mode voltage is within the acceptable range for the measurement equipment. In a measurement using an FET differential probe with a common mode voltage limit of 25 volts, a measurement is made on a digital circuit of a noise or signal voltage. It has been established that the steady state common mode voltage is within the specifications of the probe so no ground lead is used. What are the potential common mode problems for this measurement?

Answer: The difference of potential between the EUT circuit ground and the ground reference of the differential probe can easily exceed 25 volts on a transient basis due to nearby ESD events, even those too small to be noticed. In addition to affecting the accuracy of the measurement because of exceeding the common mode limits of the probe, damage to the probe itself is a real possibility unless the probe has been designed with protection against transient overload.

7

Magnetic Loop and Other Noncontact Measurements

INTRODUCTION

All of the measurement techniques discussed so far have involved direct electrical connection to the circuit under test, usually a signal and ground lead for an unbalanced measurement or two signal leads for a balanced measurement. There is a class of measurement techniques that do not involve direct electrical connection between the measurement equipment and the circuit under test, noncontact measurements. For these measurements, the test equipment is usually coupled to the circuit under test either magnetically or optically. Magnetic coupling is by far the most common coupling method used for noncontact measurements and is the subject of this chapter and Chapter 8. In this chapter, magnetic pickup loops will be discussed.

Magnetic pickup loops have many measurement uses in finding and measuring noise, as well as measuring signals, in electronic circuits. Figure 7.1 shows a few of the many types of magnetic pickup loops available. All of the loops shown are connected via coaxial cables to a measurement instrument that terminates the cable in its characteristic impedance, usually 50 ohms. The measurement instrument is typically an oscilloscope or spectrum analyzer.

Electromagnetic Compatibility, EMC, engineers have used round shielded loops for many years to "sniff out" noise sources of (magnetic)

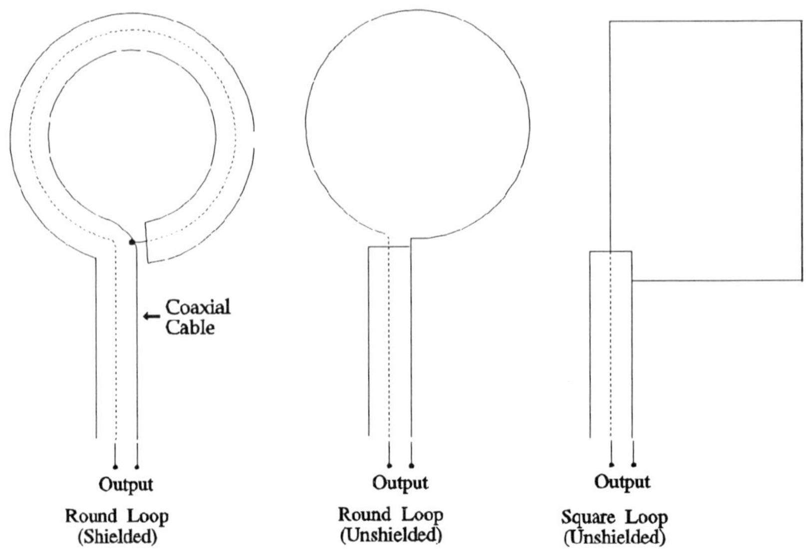

Figure 7.1 Magnetic pickup loops.

differential mode radiation[1] in digital electronic circuits. For this purpose, the loop is connected to a spectrum analyzer and passed over the circuitry under test. The spectrum analyzer displays the frequencies and relative amplitude of magnetic radiation from the circuit, allowing an engineer to pinpoint a source of noise. Some loops are calibrated so that the magnetic field intensity can be determined, but most loops used as "sniffers" are just used to provide a relative output.

Shielded loops are used to minimize stray electric field pickup, a potential problem near high electric fields. The wide dynamic range of many spectrum analyzers, on the order of 100 dB, can also make use of a shielded loop important.[2]

---

1. For more information on differential mode radiation see Noise Reduction Techniques in Electronic Systems, Henry W. Ott, Second Edition, John Wiley & Sons, pp.298-313.
2. For more information on shielded magnetic loops see Noise Reduction Techniques in Electronic Systems, Henry W. Ott, Second Edition, John Wiley & Sons, page 60.

The square unshielded pickup loop will be a main topic for this chapter. Its special shape gives it many uses which include measuring noise *voltages*, signal and noise currents, precisely locating noise sources in electronic circuits, and injecting noise into circuits to measure noise immunity. The square magnetic loop usually is not shielded for reasons that will become clear later in this chapter.

The shape of the square loop is important as it allows easy determination of the distance along which the loop has mutual inductance to a nearby circuit. Calculation of the mutual inductance of a round loop to a straight conductor is much more complicated and will be lower than that of a square loop since a round loop has no side parallel to a straight conductor.

The many uses of the square magnetic pickup loop are best understood in light of the theoretical basis for the loop's operation, the subject of the next section. Practical examples of the square magnetic loop's uses will be covered later in the chapter as will be other noncontact methods of measurement.

SQUARE MAGNETIC PICKUP LOOP — THEORY OF OPERATION

Mutual Inductance

As discussed in Chapter 2, any conductor in the vicinity of another carrying a time varying current has induced in it a series open circuit voltage, V_n, given by equation 2.9 as:

$$V_n = M \cdot (dI_1/dt) \tag{2.9}$$

where: V_n is the open circuit induced voltage in the second conductor,

M is the mutual inductance between the conductors, and

I_1 is the current flowing in the first conductor.

For the square magnetic loop, Figure 7.2 shows such a configuration.

In Figure 7.2, V_{so} is a source driving the current $i_1(t)$ around a closed loop. For the purposes of this discussion, the assumption is made that the top and bottom of the closed source loop are far enough

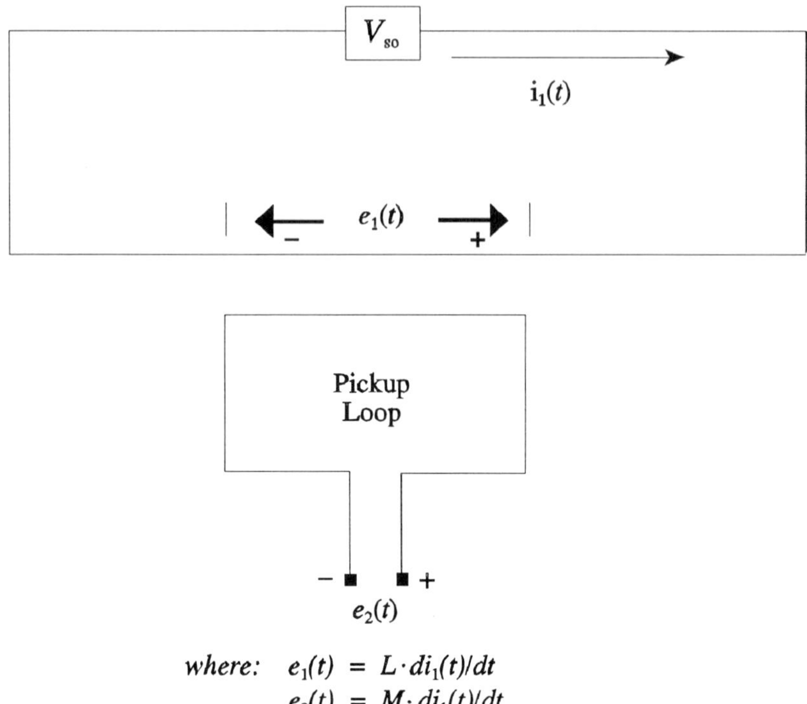

Figure 7.2 Induction into a magnetic loop.

away that all of the magnetic flux from each side is completely contained within the loop. Theoretically, this would mean that the loop would have to be infinitely large, but practically the magnetic field of a current carrying wire drops off rapidly enough so that an inch or more is almost as good. A similar assumption is made also for the pickup loop in Figure 7.2.

Current $i_1(t)$ causes a voltage drop along its path of $e_1(t)$, $L \cdot di_1(t)/dt$, where L is the inductance per unit length of the path. Since the assumption is made that the magnetic flux due to the current is contained within the closed source loop, L is not reduced by coupling from the opposite side of the loop and is on the order of 20 nh per inch. The magnetic field also induces $e_2(t)$ in the near side of the square pickup loop by the mutual inductance, M, between the two paths. Again, since the assumption is made that the opposite side of the pickup loop is far enough away so that most of the flux generated by $i_1(t)$ passes through the loop, the opposite side of the loop does not have an appreciable voltage induced in it. The other two sides of the pickup loop are perpendicular to the current carrying wire and therefore have no voltage induced in them and thus $e_2(t)$ is the output voltage of the loop.

The mutual inductance, usually measured in nanohenries, between two identical wires is less than the inductance of either wire. If the two parallel adjacent sides of the two loops in Figure 7.2 could occupy the same space, all of the flux from the section of the closed source loop next to the pickup loop will cut the pickup loop (neglecting the flux that falls beyond the far side of the pickup loop). The mutual inductance and self inductances of the wires would be the same in this case, about 20 nH per inch. However, it is physically impossible for the wires to occupy the same space and still be separate. There is always *some* space between the adjacent wires and magnetic flux will pass through it instead of through the pickup loop, reducing the mutual inductance between the two loops.

The coefficient of coupling, k, is a number that represents the fraction of flux generated by one circuit that cuts through another circuit. If two wires could occupy the same space, k would be equal to unity and the mutual inductance would be equal to the self inductances of the wires. In general, for two wires M and L are related by:

$$M = k \cdot L \tag{7.1}$$

A typical number for k between two adjacent 24 gauge hookup wires with nominal insulation is about 0.5. Thus the mutual inductance between the wires would be about $0.5 \cdot 20$ nh/in. or 10 nh/in.

Measuring Voltages Across Conductors

Since $M < L$, the open circuit voltage in Figure 7.2, $e_2(t)$, induced in the pickup loop must be less than the voltage induced across the inductance of the conductor carrying current $i_1(t)$ adjacent to the pickup loop, $e_1(t)$. Thus the magnitude of $e_2(t)$ is a lower bound for the magnitude of $e_1(t)$. Even more than that, since both M and L are real constants, the waveshape for both $e_2(t)$ and $e_1(t)$ are the same. *This means that the output of the square magnetic pickup loop is a lower bound for the voltage per unit length across the inductance of a current carrying conductor where the unit length is just the length of a side of the square loop.*

That the square magnetic pickup loop can measure a voltage in a circuit without making direct electrical contact with the circuit makes it a powerful tool for measuring noise voltages. If a round loop were to be used for this purpose, the determination of the distance over which the mutual induction between the loop and the circuit under test is significant, although it can be calculated, and is much more complicated than for the square loop. Near the end of this chapter, practical examples on how to make use of this tool will be discussed.

Signal voltages are generally not measured across the inductance of conductors. In addition, many signal measurements require more amplitude accuracy than that afforded by the uncertainty of the coefficient of coupling between the pickup loop and the circuit under test. These characteristics make the square loop less useful for signal measurements.

FACTORS AFFECTING THE SIZE AND SHAPE OF THE PICKUP LOOP

Ease of Use

There are several factors that determine the exact size and shape of the pickup loop that works best in a given measurement application. One of the most important is ease of use. To measure voltages induced on printed wiring board traces, the loop needs to fit in between components and also be small enough to localize a measurement to a relatively small section of a path. However, the loop needs to be large enough to pick up a useable signal. A 1 inch square loop seems to be a reasonable compromise between sensitivity and ease of use. One could reduce the sides of the loop parallel to the circuit under test to 0.5 inch to increase resolution on the circuit under test, but the sides of the loop perpendicular to the circuit under test should not be made smaller than 1 inch. Distances smaller than this will allow too much magnetic flux from the circuit under test to fall

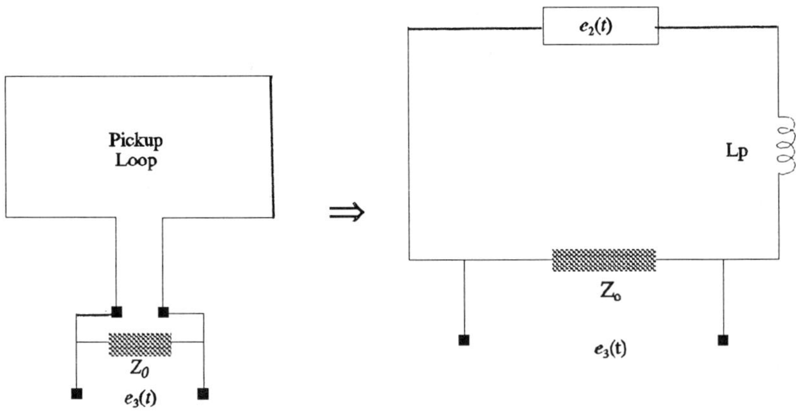

Figure 7.3 Pickup loop equivalent circuit.

where: $e_2(t)$ is the open circuit loop voltage, and
  $e_3(t)$ is the loaded loop output voltage.

outside of the pickup loop. A 1 by ½ inch rectangular loop would work well in this case.

Sensitivity vs. Frequency Response Tradeoff

The open circuit voltage generated in the pickup loop was given in Figure 7.2 as:

$$e_2(t) = M \cdot di_1(t)/dt. \tag{7.2}$$

This open circuit voltage is in series with the pickup loop self inductance, L_p. Since the loop is loaded by a coaxial cable terminated in its characteristic impedance, Z_0, a voltage divider is formed by these two elements which reduce the output of the loop. The equivalent circuit for the pickup loop is shown in Figure 7.3. The loop output voltage $e_3(t)$ is given by:

$$e_3(t) = e_2(t) \cdot \frac{Z_0}{Z_0 + j\omega L_p} = e_2(t) \cdot \frac{1}{1 + j\omega L_p/Z_0}. \tag{7.3}$$

Remember that $e_2(t)$ is a lower bound for the voltage across the conductor adjacent to the pickup loop.

From Equation 7.3 it is apparent that the load on the loop's output terminals combined with the loop's self inductance form a low pass filter that reduces the loop sensitivity above a cutoff frequency equal to $Z_0/2\pi L_p$. Thus if a larger or multiturn loop is used to increase the mutual inductance, M, the loop self inductance, Lp, is also increased and the cutoff frequency is reduced. A single turn one inch square loop is a good compromise between frequency response and sensitivity.

EXAMPLE

7.1   A single turn, 1 inch square loop will have a self inductance of about 80 nh (20 nh per side). If connected to a 50 ohm coaxial cable, the cutoff frequency given by Equation 7.2 is about 100 MHz. After that frequency, the loop voltage response falls off at 6 dB per octave of frequency (20 dB per decade).

The voltage response of a pickup loop is shown in Figure 7.4. The vertical scale is proportional to base 10 log of the loop output voltage for a constant value of induced voltage across the adjacent

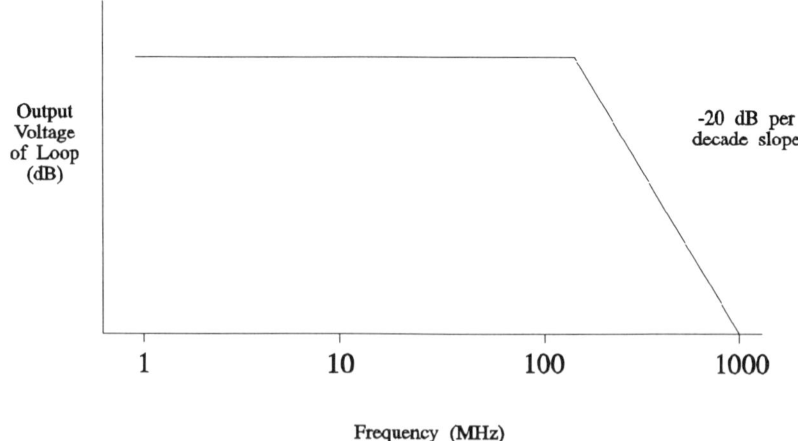

Figure 7.4 Voltage response of a pickup loop.

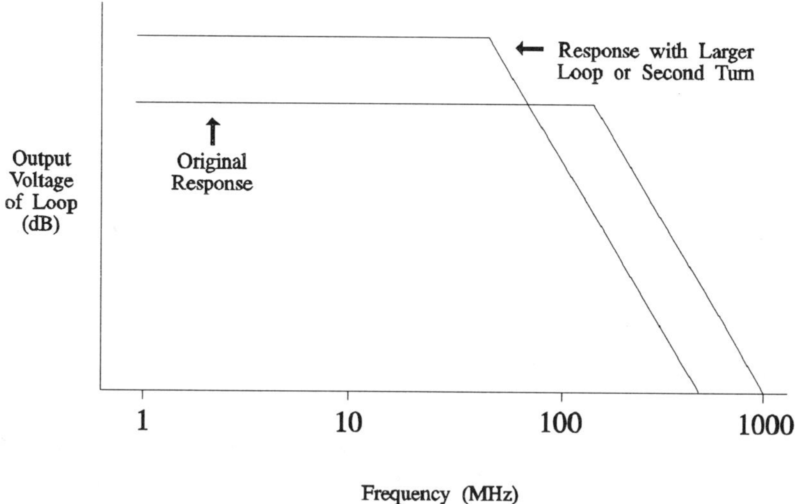

Figure 7.5 Voltage response of a pickup loop
(larger loop or with added turns).

wire being measured. The horizontal scale is a log frequency scale.
The corner frequency shown is typical for a 1 inch square loop.

The output voltage is a constant fraction of the voltage across the
wire being measured until the frequency exceeds the cutoff frequency (100 MHz for a 1 inch square loop). Above that frequency,
the loop's output voltage falls off as a single pole response, that is at
-20 dB/decade. The loop is still useful above the cutoff frequency, it
just becomes less sensitive.

Enlarging the loop or adding a second turn raises the flat portion
of the curve in Figure 7.4, but also lowers the corner frequency, and
thus its useful bandwidth for voltage measurements, as shown in
Figure 7.5. Adding a second turn can compromise the loop's performance further by adding interturn capacitance that can cause
resonance conditions in the loop and alter the frequency response of
the loop from that shown in the figure. A typical 1 inch square, one
turn loop will follow the curve shown in Figure 7.4 up to just above
1 GHz before deviating from the −20 dB per decade slope. A
multiturn loop's frequency response will typically show significant
deviations well below 1 GHz, limiting its useful frequency range.

Orientation of Loop

Often the pickup loop will be used to measure noise voltages on Printed Wiring Board, PWB, traces, such as a power or ground trace. The loop should be held perpendicular to the circuit board to prevent unintended induction into the three sides of the loop not adjacent to the trace being measured. The loop construction should be such that the loop can be easily held perpendicular to the PWB with the cable exiting the loop also perpendicular to the board.

CURRENT RESPONSE OF THE PICKUP LOOP

A square pickup loop can also be used to estimate the current flowing in an adjacent wire. There are many applications where current measurements are important. Several of these applications are covered in the following chapters of this book. Using a pickup loop to estimate high frequency current flowing in a conductor is an inexpensive method that often yields acceptable results.

Figures 7.4 and 7.5 showed the output voltage of a pickup loop as a function of frequency for a constant voltage across the unit length of conductor being measured. This is useful when using an oscilloscope to measure the peak voltage across a ground path, for instance, in the time domain. To determine the time domain current one would have to integrate the voltage waveform displayed by the oscilloscope. Integrating Equation 2.9 yields:

$$V_n = M \cdot (dI_1/dt) \tag{2.9}$$

$$\int V_n \cdot dt = M\, I_1 \tag{7.4}$$

where: V_n is the open circuit induced voltage in the second conductor,

M is the mutual inductance between the conductors, and

I_1 is the current flowing in the first conductor.

In many instances, it is desired to measure the current in a conductor in the frequency domain, usually with a spectrum analyzer. This type of measurement is important in predicting potential radiation due to common mode current on a cable. This topic is discussed in Chapter 8.

In the frequency domain, the derivative of a single frequency source is simply $j \cdot \omega$ times the original source as shown in equation 7.6.

$$I_1 = I \cdot e^{j\omega t} \qquad (7.5)$$

$$dI_1/dt = j\omega \cdot I \cdot e^{j\omega t} = j\omega I_1 \qquad (7.6)$$

Thus Equation 2.9 becomes:

$$V_n = M \cdot (dI_1/dt) \qquad (2.9)$$

$$V_n = M \cdot j\omega I_1. \qquad (7.7)$$

From Equation 7.7, it is seen that for a constant current amplitude in the wire being measured, the open circuit output voltage of the loop increases linearly with frequency.

The loaded loop output voltage was given by Equation 7.3 as:

$$e_3\,(t) = e_2\,(t) \cdot \frac{Z_o}{Z_o + j\omega L_p} = e_2\,(t) \cdot \frac{1}{1 + j\omega L_p/Z_o}. \qquad (7.3)$$

Combining Equations 7.3 and 7.7, noting that V_n and $e_2(t)$ both represent the open circuit voltage of the loop, the loaded loop output voltage when held next to a wire carrying a constant current is:

$$e_3\,(t) = M \cdot I_1 \cdot \frac{j\omega}{1 + j\omega L_p/Z_o}. \qquad (7.8)$$

The output voltage of the loop can be seen to be the product of a term that increases linearly with frequency (the numerator of Equation 7.8) at a rate of 20 dB per decade and a term that is flat to a corner frequency and then decreases at 20 dB per decade above that frequency (the denominator of Equation 7.8), the same as the voltage response of the loop discussed earlier.

An intuitive way to look at the loop's current response is the following. The voltage across the length of the wire being measured by the loop for a constant current increases linearly with frequency, because the inductive reactance of the wire increases with frequency. So the open circuit voltage induced into the loop by a constant current increases with frequency at a rate of 20 dB per decade. But, the voltage response of the loop falls off after its cutoff frequency by precisely the same amount, 20 dB per decade. Therefore the loop output voltage for a constant current rises at 20 dB per decade of frequency until the cutoff frequency after which it remains constant with frequency as the increasing induced open circuit voltage is cancelled by the decreasing frequency response of the loop. The situation is shown graphically in Figure 7.6 for a 1 inch square loop having a cutoff frequency of 100 MHz. The dotted line indicates the response of the loop, the total of the two components. The loop current response for a 1 inch square loop is typically flat from about 100 MHz to beyond 1 GHz.

PICKUP LOOP NULL EXPERIMENTS

A square pickup loop is usually unshielded. This allows the loop to get closer to a current carrying wire for a measurement. Since mutual inductance falls off rapidly with distance from the loop to the current carrying wire under test, a tenth of an inch spacing will noticeably

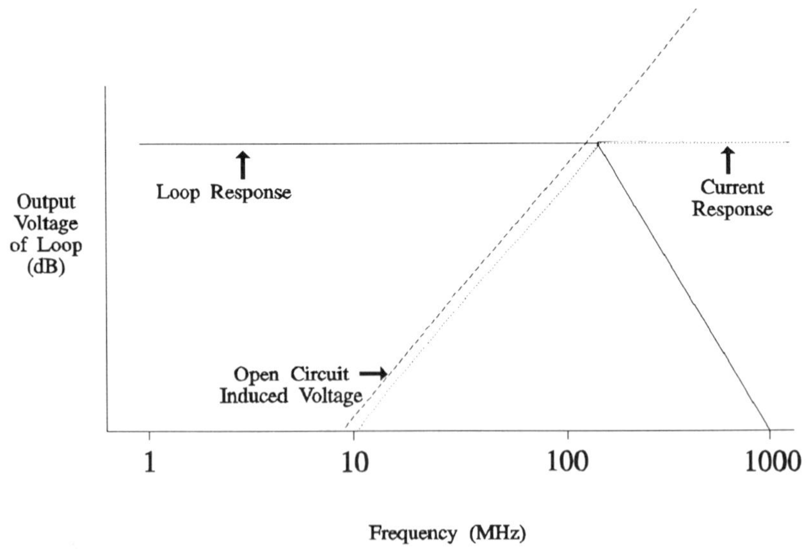

Figure 7.6 Current response of a pickup loop.

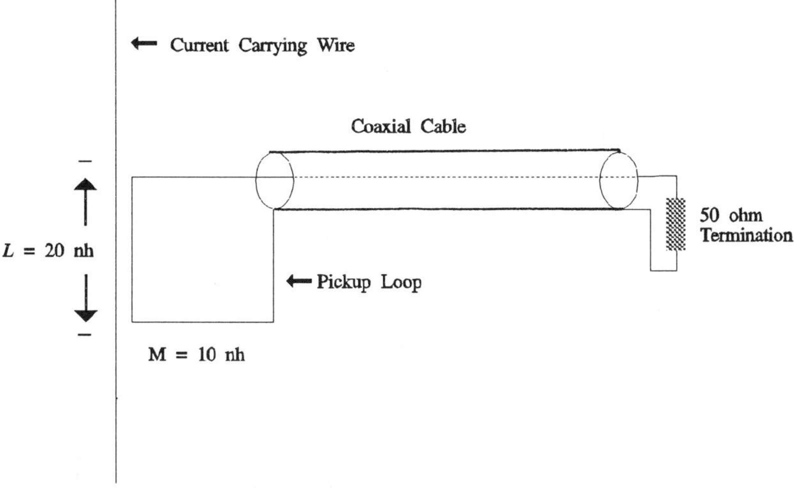

Figure 7.7 Pickup loop effect on a circuit under test.

reduce the loop's output. However an unshielded loop can be subject to electric field pickup near rapidly changing strong electric fields. In order to be certain that the loop output reflects magnetic induction from the circuit under test, a null experiment should be performed to validate the measurement.

Null experiments were introduced in Chapter 4 as means of validating a measurement. A null experiment is one which should have a known (usually trivial or zero, hence the name) outcome. For the pickup loop, one such null experiment is simply to rotate the loop 180 degrees while holding it against the circuit under test. If the pickup loop output is caused by magnetic coupling to the circuit under test, it will drop to zero when the loop is turned 90 degrees to the circuit under test and the waveform displayed on an oscilloscope will invert as the loop passes through 90 degrees. Be careful. If the displayed waveform is symmetrical, such as a sinewave, it may look the same when inverted because of the way scopes trigger on a waveform, that is, on a level and slope of the input waveform. When using a spectrum analyzer, the phase difference cannot be observed, but the output should null when the loop is turned 90 degrees to the circuit under test.

Another null experiment involves just pulling the loop away from the circuit under test. Even a distance of one quarter inch will lower the loop output significantly, as long as its output is not caused by stray magnetic fields.

EFFECT OF THE PICKUP LOOP ON CIRCUIT OPERATION

Since the pickup loop does not directly contact the circuit under test, it has much less effect on circuit operation than measurement methods requiring direct electrical contact to the circuit under test, such as a voltage measurement using a scope probe. But, what effect does it have? The effect can be calculated from the diagram in Figure 7.7. The figure shows a 1 inch square loop positioned next to and parallel to a current carrying wire. Along the inch of wire next to the loop, the wire would have about 20 nh of inductance and the wire and loop might be coupled by about 10 nh of mutual inductance assuming k, the coefficient of coupling between the wire and loop to be about 0.5 (a realistic number). The equivalent circuit of Figure 7.7, neglecting the isolation between the loop and wire, is shown in Figure 7.8.

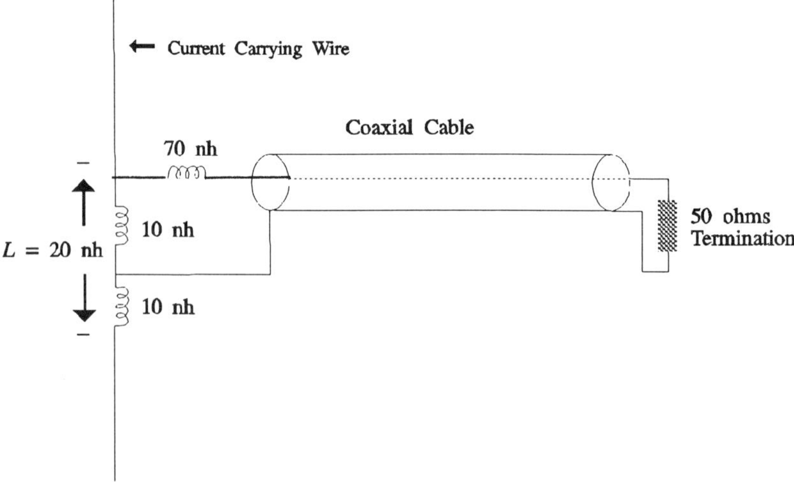

Figure 7.8 Equivalent circuit of a pickup loop and a circuit under test.

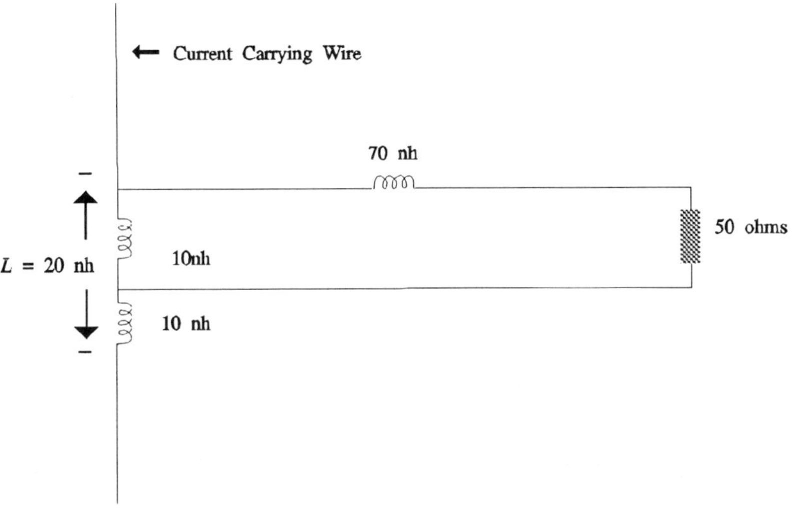

Figure 7.9 Equivalent circuit of a pickup loop and a circuit under test.

The wire's 20 nh of inductance can be split into two 10 nh inductances, only one of which is coupled to the measurement circuit (remember the isolation between the loop and wire is being neglected for this discussion). Thus even if the measurement circuit were a short circuit, the only effect on the circuit is to reduce the wire under test inductance by 10 nh, about the equivalent of one half inch of length. But, the effect of the pickup loop is even less than that. The other three sides of the pickup loop have about 60 nh of inductance. That inductance in addition to the remaining 10 nh of the pickup loop self inductance not coupled to the current carrying wire can be represented by a discrete 70 nh inductor shown in Figure 7.8, and that is in series with the input impedance of the coaxial cable, Z_0. The equivalent circuit becomes that shown in Figure 7.9. So, a 1 inch square pickup loop appears as a 70 nh inductor in series with 50 ohms and that combination is across only 10 nh of the wire under test, a very small effect indeed. The only other loading of the circuit under test is that of parasitic capacitance between the loop and circuit under test. This parasitic capacitance, on the order of a few pF at most, is unlikely to have any measurable effect on the circuit except at very high frequencies (above several hundred MHz). At those frequencies, the wire under test would not likely be an exposed wire (unless it were part of an intentional antenna) and therefore not a candidate for this measurement.

THE PICKUP LOOP TECHNIQUE OF LOCATING NOISE SOURCES

In order to measure and mitigate the effect of a noise source, one must first find the noise source. Using traditional methods, this can be a time and resource consuming job. For many situations the use of a *pair* of pickup loops can find problem noise sources in electronic systems quickly and easily. Why use two loops? By using two loops, the relative *phase* of noise currents can be determined as well as the

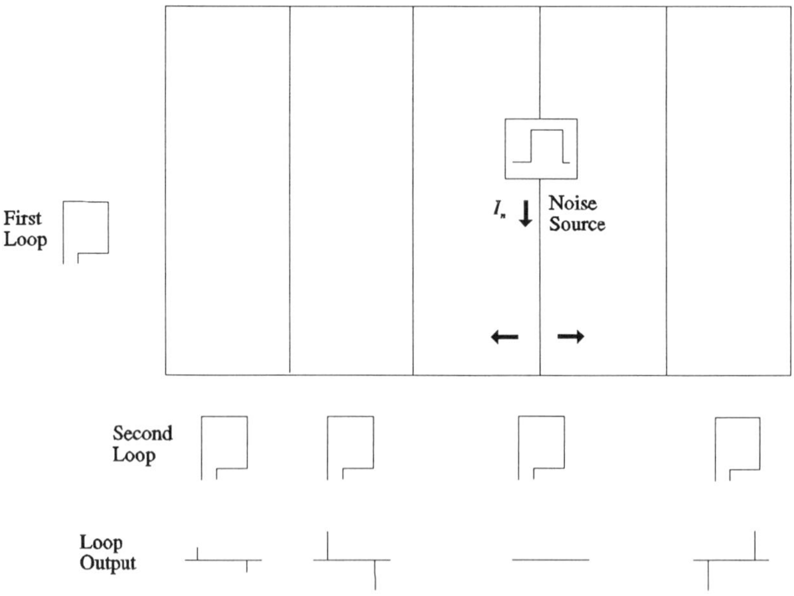

Figure 7.10 Finding a noise source in a grid of conductors using two pickup loops.

amplitude, and it is the phase information that allows easy location of the noise sources.

The procedure is as follows. Use one pickup loop to display the noise signal in an accessible part of the system or circuit caused by a noise source at an unknown location. This loop is connected to one vertical input of an oscilloscope and the sweep is triggered from this signal. This gives the scope a stable trigger and eliminates triggering problems that can otherwise affect the displayed signals from a second moveable loop. The second loop, connected to another vertical input of the oscilloscope, is then moved around the system or circuit. Using the relative phase between the two oscilloscope traces, the direction of the noise currents in the system can be mapped out as well as their amplitude. With this information the noise source can usually be easily determined. Figure 7.10 illustrates this process for a simple case.

In Figure 7.10, a noise source is shown located in one branch of the circuit composed of several branches. This circuit could represent many circuits from the small ground grid of a PWB to the rather large circuit of the 120 VAC powerline distribution to several pieces of interconnected equipment. The noise current, I_n, flows down the branch containing the noise source and splits at the bottom where some of the current travels to the left and some to the right.

The first loop is positioned on the leftmost path of the grid and picks up a small signal caused by the noise source. This signal displayed on an oscilloscope provides a stable trigger and a phase reference for the second moveable loop.

The second loop is initially positioned on the same branch as the first loop and will display the same waveform, shown at the bottom of the figure. The waveform for this case will be *di/dt* spikes corresponding to the edges of the pulse generator shown as the noise source. As the loop is moved toward the branch containing the noise source, the magnitude of its output signal will increase. This arises because of the fact that each branch carries some of the noise current and as the second loop gets closer to the source branch, its sensed current includes the current flowing in the branches already passed on the left.

A dramatic occurrence happens as the loop is passed by the branch containing the noise source. When the loop is directly under the noise source branch it senses some of the noise current flowing to the left and some of it flowing to the right. These two components of noise current produce cancelling signals in the loop, since the sensed currents are in opposite directions, and the loop output drops to a null, zero output. Once the second loop passes to the right of the noise source branch, it senses current flowing to the right instead of the left and its output waveform *inverts* as shown in Figure 7.10. Without the stable trigger provided by the first loop in this example, the waveform inversion in the second loop output would not appear to happen because the scope would still trigger on the same polarity edge of the second loop.

The stable trigger can be provided by other means than a second loop. A scope probe measuring the noise voltage at a point in the circuit could be used to provide the trigger, although with the disadvantage of a direct connection to the circuit. In general, any means of displaying the noise at some point in the circuit on an oscilloscope can be used to provide the stable trigger so that the moveable loop yields phase as well as amplitude information.

Figure 7.11 shows three typical cases of current splitting at nodes. In Case 1, current I_1 flows down into the node and splits into I_2 to the left and I_3 to the right as was the case in Figure 7.10. The signal from a pickup loop passing from left to right along the bottom trace will show a null and phase inversion as the node is passed, as in Figure 7.10.

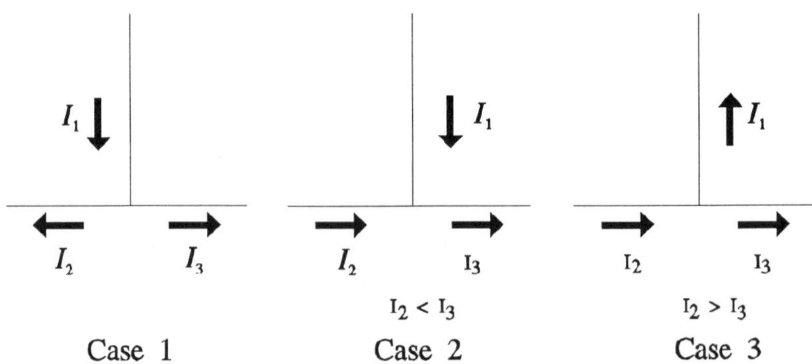

Figure 7.11 Types of current nodes.

Case 2 has a current entering from the left, I_2, joining in phase with a current coming in from the top, I_1, to form the larger current I_3. As a loop is passed along the bottom path from left to right, its output signal will increase as it passes under the node to reflect the sum of I_1 and I_2. There will be no phase inversion however since I_2 and I_3 have the same relative direction.

Case 3 shows what happens when the direction of I_1 is reversed such that it subtracts from I_2. I_3 will be reduced but retain the same phase as I_2 as far as a pickup loop passing from left to right is concerned.

The above three examples show how the current patterns flowing in a circuit can be determined by using the two loop technique. In all cases, the current leaving a node must be equal to the current entering the node according to Kirchoff's current law. There are many other

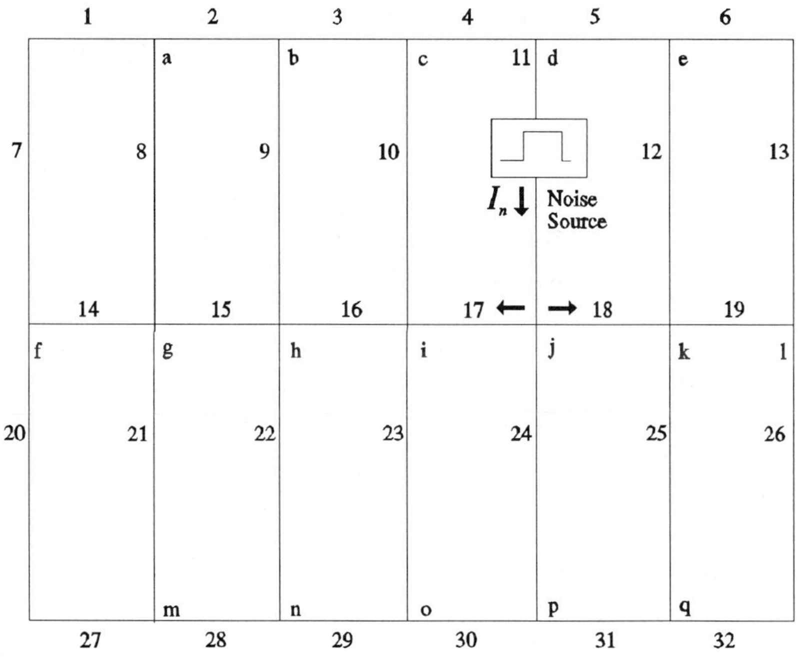

Figure 7.12 Finding a noise source in a complex grid of conductors using two pickup loops.

configurations that can be analyzed by applying Kirchoff's current law to locate a source of noise in a circuit.

Let us analyze the more complex circuit shown in Figure 7.12. Each of the branches is numbered and the nodes are marked by letters. The noise source, as before, is located in one of the branches. The pattern shown in Figure 7.12 could be a ground grid on a circuit board or, on a larger scale, could represent the grounding system of a large multicabinet electronic system such as a mainframe computer or a telephone central office switching system.

The noise source of Figure 7.12 is located in branch 11 and produces noise current I_n that flows into node j and divides up into branches 17, 18, and 24. After obtaining a stable trigger from a stationary loop or scope probe, a moveable loop can be then used to track down the noise source.

As the moveable loop (starting at node f) is passed from left to right along branches 14, 15, 16, and 17, the displayed signal on the scope grows larger as the contributions from branches 8, 9, 10, 21, 22, and 23 are added into the total current seen by the pickup loop. A null and phase inversion will occur as the loop passes node j just as for the simpler case of Figure 7.10. Since branches 24 and 11 are perpendicular to the loop they do not affect the null as the loop passes node j. Similarly, as the moveable loop is passed from left to right along branches 27 through 32, the amplitude of the loop output, although smaller, will increase, pass through a null at node p, and then undergo a phase inversion to the right of node p.

In fact, using the moveable loop with a stable trigger, one can map out the relative direction and amplitude of the current in each branch of the circuit. Some of the branches may have a zero current for this example. Likely candidates are branches 22 and 25. This happens because the voltage across the branch would be zero if the branch were not there, so when it is added no current will flow.

The assumption has been made in the above example that the branches of the grid are a small fraction of a wavelength for the highest frequency involved. For circuits that do not meet this condition, an extra phase shift will occur in the currents and voltages in the grid. However, at the junction of the branch containing the noise source, a null and phase reversal will likely still be evident.

OTHER NONCONTACT MEASUREMENTS

General

Up to this point, all of the measurement techniques discussed in this book have been direct, that is, a measurement of a specific voltage or current at a known point in the circuit. In this chapter, the square magnetic loop and its measurement potential have been discussed. One is making a *noncontact* direct measurement of the desired voltage when using the square loop in the fashion described earlier. There are other methods of making measurements where the test equipment is not conductively coupled to the circuit under test. A couple of these are discussed below.

Current Probes

Current probes have a voltage output that is proportional to the current flowing through them. They in effect measure the magnetic field produced by the current to be measured. Earlier in this chapter it was shown how the square magnetic loop can be used for this purpose. One drawback of using the square magnetic loop for current measurements is the unknown coefficient of coupling, k, between the loop and the circuit under test. Due to varying insulation thickness and the need for the wire to be measured to be precisely parallel to one edge of the loop, measurement amplitude error of ± 50 percent is possible.

Current probes with calibrated response over a wide frequency range are available commercially. To measure a current in a conductor, it is only necessary to pass the conductor through the center of the probe and measure its output voltage, usually with an oscilloscope or spectrum analyzer. Chapter 8 will discuss how current probes work and give some typical applications, but for now it is sufficient to say that current probes represent one of the most useful measurement tools available for measuring signals and noise in electronic circuits.

Optically Coupled Measurements

Optically coupled measurements involve a voltage or current to optical transducer, some length of optical fiber, and an optical to electrical conversion at the measurement equipment end. Measurements of this type provide very high isolation between the equipment under test and the measurement equipment. In addition, the measurement equipment can be located remotely at a relatively great distance from the equipment under test. These characteristics make optical measurements well suited for measuring the effects of severe interference on equipment.

Optically coupled measurements have been used in measurements of spacecraft response to ESD for years.[3] The electrical to optical convertors are small battery powered boxes that are located in the spacecraft to make measurements in critical locations. Often the optical fibers will be run a couple of hundred feet to the remote measurement equipment where the optical to electrical conversion is done. Measurements made in this way minimize the affect of the measurement equipment on the spacecraft while minimizing the affect of the ESD on the measurement equipment. Chapter 9 will cover in detail techniques for making measurements in the face of severe interference such as ESD.

Since optical measurements are often used near severe interference, performance of null experiments is a must to insure that the measurement setup, especially the electrical to optical conversion, is not susceptible to the interference.

Indirect Measurements

Sometimes a measurement needs only to yield limited information about a signal or noise voltage or current. For example, the noise

---

3.   For more information see: "Electrostatic Discharge (ESD) Response Measurements in a Satellite Environment", G. H. Tucker, 1987 *EOS/ESD Symposium Proceedings,* EOS-9, pp. 124-128.

margin of a digital bus, the difference between the noise on the bus and the logic thresholds of the bus receivers. In this measurement, it is only necessary to measure the peak value of the noise voltage that occurs on the bus. It is generally *not* necessary to know the exact waveform or the location along the bus of the noise (except possibly for trouble shooting purposes).

Indirect measurements fill such a need. In an indirect measurement, limited information about a voltage or current in the circuit under test is inferred from experiments on the circuit under test. In general, an indirect measurement must be tailored to the equipment under test and requires detailed knowledge about how the equipment under test works.

EXAMPLE

7.2:   Consider the measurement of the noise margin of a digital bus that uses current mode bus drivers. Each driver sends a measured amount of current to the bus for a logical "one" and no current for a logical "zero" or when it is not enabled.

Since the bus drivers are current sources, in this example, one can apply a DC bias to the bus to force a DC offset without adversely affecting the bus drivers. This characteristic provides one with a way to measure the bus noise margin *for this case only*. If the bus drivers were low impedance voltage drivers, a more common type, trying to force a DC offset on the bus would only heat up the bus drivers that were active.

To determine the noise margin of the bus for this example, connect a variable DC voltage source through a resistance much higher than the characteristic impedance of the bus. A 50 ohm bus should not be affected by using 2000 ohms for this purpose.

Adjust the DC offset in a positive direction above ground slowly until the bus malfunctions. The amount of DC offset required to do this is the noise margin of the bus since the DC offset plus the noise in the system must be adding up to the bus receiver threshold when the malfunction occurs.

Noise Injection

One type of indirect measurement is to inject noise into a circuit until a malfunction occurs. The amount of noise injected to cause the malfunction is the noise margin of the circuit. Probably the most convenient way to inject noise on a printed wiring board trace is with the square magnetic loop. Just as the loop can measure a voltage per unit length across a circuit under test, it can also inject a voltage per unit length into a circuit under test just by holding the loop close and parallel to the conductor under test. It is only necessary to use a source with a high enough di/dt to drive the loop to inject the required noise into the circuit.

There are also commercially available injection probes that are similar in construction to current probes. These types of probes are generally best suited for use on cables as opposed to printed wiring paths. To use one of these probes on a printed wiring path, one would need to cut the path and install a jumper long enough to pass through the probe. Because they are often designed to handle considerable amounts of power, size and weight are also a problem for printed wiring board use.

For logic circuits a few volts of noise will generally do. Unfortunately, a normal logic pulse generator generally does not have a high enough di/dt for this purpose. A more powerful generator such as the Hewlett Packard 214B is necessary. For higher amounts of injected noise, one should consider the burst generator used to determine compliance with the IEC 801-4 Electrical Fast Transient Test.[4] The 214B generator can induce as much as 10 volts of peak noise into a circuit via a 1 inch square loop.

There are several factors which can increase the accuracy of this type of test. First, the amplitude and pulse shape of the injected noise should be monitored with the balanced coaxial probe described in

---

4.   IEC 801-4 is an international standard for measuring the noise immunity of equipment to the type of high frequency noise generated by arcing switch contacts. A copy of the standard is available for a charge from the American National Standards Institute, 1430 Broadway, New York, NY 10018.

Chapter 6. The probe should be connected to the receiving end of the path. Second, since an error condition usually relies on the coincidence of an injected pulse and a condition of a signal in the circuit under test, such as a clock edge, the amplitude should be raised from zero through the expected noise margin *slowly* taking a couple of minutes. Make sure the noise generator is not set to the system clock frequency or any multiple of it.

Not only is this method of noise margin measurement fast, but it is extremely easy as well. Most of the time it involves holding the square loop close to the circuit under test and connecting a balanced coaxial probe to monitor the resultant noise injection. Sometimes it is the only method of noise margin measurement that can be practically used. As with most indirect measurements, only the amplitude of the noise margin can be determined. No information is available on the waveshape of the internal circuit noise sources or their location.

EXAMPLE

7.3:   Consider the measurement of the noise threshold level of an Address Latch Enable, ALE, lead in a microprocessor circuit using noise induction. Noise is injected by holding one edge of the square loop in contact with the ALE lead while the amount of noise injected is monitored at the memory with a balanced coaxial probe, Chapter 6, using no probe ground lead.

Setting the generator to deliver $L \cdot di/dt$ spikes at a few MHz, but not at the system clock frequency, its amplitude is slowly increased. The peak reading from the monitor probe at the time of circuit malfunction should be close to the memory's dynamic noise margin.

SUMMARY

This chapter has covered noncontact measurement techniques and magnetic loops. In particular, the characteristics and advantages of the square magnetic loop in circuit measurements and troubleshooting were discussed along with other noncontact and indirect measurement methods.

The square loop can be used to measure a lower bound for the voltage induced per unit length across the inductance of a circuit under test. The unit length for the measurement is the length of the loop along the circuit under test. For most purposes the square loop should have only one turn and be connected to 50 ohm coaxial cable properly terminated at its other end.

Additional turns on the loop increase its sensitivity below its corner frequency and lowers its sensitivity above that corner frequency, the net effect being to give higher sensitivity but with a flat frequency response to a lower frequency. Usually more turns will compromise the frequency response at the higher frequencies. For a one turn 1 inch square loop, its corner frequency is about 100 MHz. Above that voltage response falls off.

The square magnetic loop also has an output proportional to current flowing in the circuit under test. For a 1 inch square loop, its response to current is flat from about 100 MHz to over 1 GHz. Due to variations in the coefficient of coupling, k, to the circuit under test, measurements of current are qualitative rather than quantitative. It is still useful for some current measurements as will be discussed in Chapter 8.

Perhaps the best use of the square magnetic loop is using phase information between two loops to locate a noise source in a circuit. An example was given to illustrate the method. As the movable loop passes the branch containing the noise source, its output will pass through a null and will invert with respect to a fixed triggering loop elsewhere in the circuit.

Other important noncontact and indirect measurement techniques were also discussed including optical, current probe, bias injection, and noise injection. Each of these methods has unique advantages in certain measurement situations. Indirect measurements generally yield information only about the peak amplitude of a noise and little if any information about the waveshape of it. Current probes will be discussed in greater detail in Chapter 8.

Each of the measurement techniques discussed in this chapter represent a powerful, if not a little unusual, tool for investigating electronic circuits. Many of these techniques have successfully solved design and interference problems in minutes or hours where conventional techniques failed after weeks or more of work. The square magnetic loop is one of the most potent tools available for noise investigations.

LABORATORY DEMONSTRATIONS

EXPERIMENTS

Equipment needed:

- 5-50 MHz square wave oscillator using HC240 inverting buffer as described in Chapter 3.
- Dual channel analog or digital scope with at least 100 MHz bandwidth.
- Balanced coaxial probe as described in Chapter 6.
- Assortment of jumpers and cables.

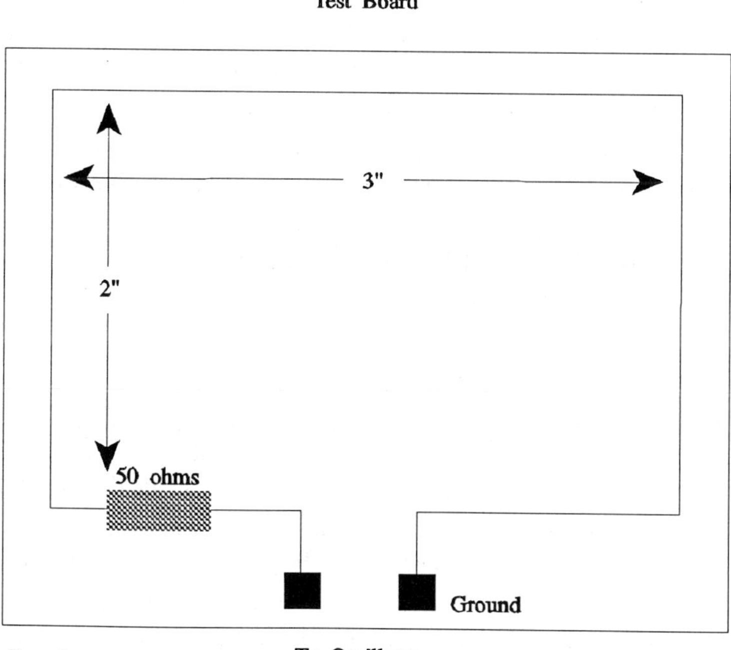

Figure for Example 7.1   Setup for Experiment 7.1.

Experiment 7.1: Typical Square Magnetic Loop Output Voltages

1. Connect the output of the 5-50 MHz oscillator through a 50 ohm coaxial cable of 6 feet or less in length to a small circuit board which contains a series 50 ohm resistor (to protect the oscillator output) and a 6 inch length of bus bar wire in series. The wire should be arranged as shown in Figure Ex7.1.
2. Set the oscillator to 5 MHz and connect to the circuit board.
3. Connect a 1 inch square loop through a length of 50 ohm coaxial cable to a vertical input of the scope and hold the loop up to the path on the circuit board. Be sure to terminate the cable in 50 ohms at the scope. It is best to use a scope with a built 50 ohm termination, but it is acceptable to use a one megohm input with an external 50 ohm load. Set the scope to trigger on the positive edge of the input signal and the sweep to about 200 ns per division so that a couple of complete cycles of the signal can be viewed. Note how the waveform looks and its amplitude.
4. Turn the loop 180 degrees still holding it against the path on the circuit board. Note what happens to the displayed waveform.
5. Connect a second loop to the second vertical amplifier input, display its output also on the scope screen, and hold it up to the path. Its output should look nearly identical to the first loop. Still triggering on the first loop, rotate the second loop 180 degrees again and note what happens to its output.

Questions:

7.1.1 What does the waveform displayed look like in step 3?

Answer: The displayed waveform will be the differential of the current, $M \cdot di/dt$. Since the current waveform is close to a squarewave, the displayed signal will be alternating positive and negative spikes. The amplitude of the spikes will be on the order of 100 mV if the HC240 based oscillator is used as a signal source. This represents a significant induced voltage caused by a buffer that is not particularly fast (about 3 ns rise and fall times) or powerful (a few tens of ma of available current).

7.1.2   What happens to the waveform in step 4?

Answer: Since the scope is set to trigger on a positive edge, a positive *di/dt* spike will be the first feature of the waveform for both orientations of the loop and both waveforms will look identical.

7.1.3   How does the waveform in step 5 compare to that in step 4?

Answer: Since the induced voltage in the second loop will invert polarity when the loop is rotated and since the scope is triggered from the first loop, the second loop's output will invert with respect to the first loop as the second loop is rotated 180 degrees. Using one loop as a stable triggering source so that the phase relative to a second loop can be determined is a powerful tool for tracing the paths of signal and noise currents in electronic circuits.

Experiment 7.2: Checking Square Loop Output with a Balanced Coaxial Probe

1.   Using the balanced coaxial probe described in Chapter 6, measure the induced voltage across 1 inch of the path in Figure Ex7.1 above. Note the amplitude and waveshape of the displayed signal. Be sure to keep the leads of the probe as close to perpendicular as possible to the signal path on the board to prevent induction into the probe leads.
2.   Make the measurement at several points along straight sections of the path and note the signal characteristics displayed on the scope.

Questions:

7.2.1   How does the displayed signal compare with the signal picked up by the 1 inch square loop in Experiment 7.1?

Answer: The waveshape should be the same and the amplitude should be somewhat higher since the balanced probe is sensing $L \cdot di/dt$ and the pickup loop $M \cdot di/dt$.

7.2.2   How does the inductive drop change along the path?

Answer: It should not change substantially because the path length is small compared to a wavelength of the highest frequencies involved. Therefore, the total path induced voltage drop is just the sum of each of the parts of the loop.

Experiment 7.3: Source Location Using Two Loops and Circuit Resonance

1.   Build a grid pattern on a breadboard using the pattern of Figure 7.12 adding a .01 µf bypass capacitor in each branch of the circuit as shown below in Figure Ex7.3. The vertical branches, such as branches 8 and 20, should be about 1.5 to 2 inches in length and the horizontal branches, such as branches 14 and 15, should be about 1 inch in length.

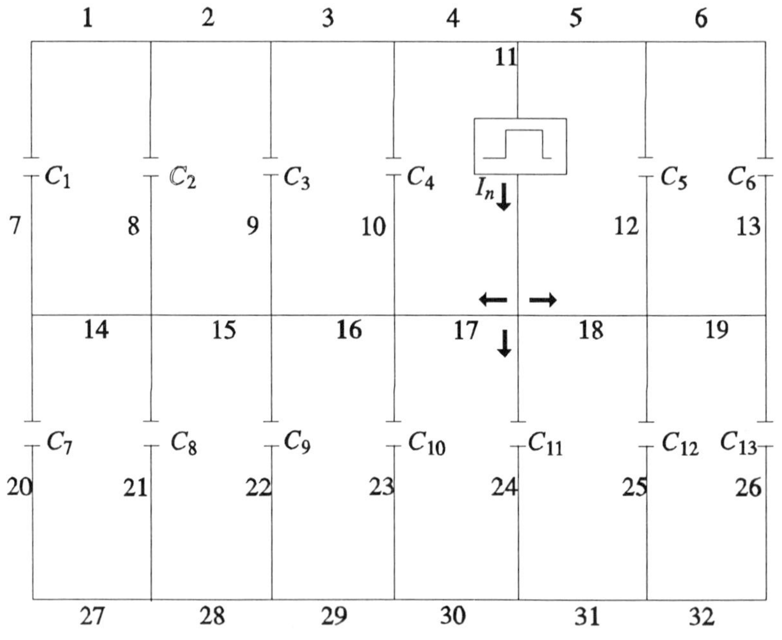

Figure for Experiment 7.3   A complex grid of conductors with bypass capacitors.

2. Connect the 5-50 MHz oscillator in series with a 47 ohm carbon resistor into branch 11 as shown. Connect the trigger input of the scope to the generator or trigger the scope from a fixed loop taped to branch 13. Adjust the oscillator between 5 and 10 MHz while holding a movable 1 inch square pickup loop next to branch 12 and note how the waveshape varies. It is composed of a sinusoidal component and an $L \cdot di/dt$ component, spikes due to the edges of the square wave excitation. Adjust the frequency for the maximum amplitude of the sinusoidal component.

3. Move the loop between branch 12 and branch 13 noting how the displayed waveshape varies.

4. Now lay the loop down so that three of its sides are parallel to branches 12, 13, and 19. Note the waveshape displayed on the scope.

5. Repeat steps 2, 3, and 4 using branches 7, 8, and 14.

Questions:

7.3.1 What causes the sinusoidal component of current in the grid?

Answer: The conductors on the board are essentially inductors above the audio band and these inductors resonate with the capacitors installed in each branch. For this circuit, the resonant frequencies are about 5 to 7 MHz. Resonant frequencies on this order or even lower are common in the power/ground grid on two sided circuit boards.

7.3.2 What is the relationship between the sinusoidal current and the $L \cdot di/dt$ generated spikes in branch 12 compared to branch 13? Why?

Answer: The phase of the $L \cdot di/dt$ spikes remains the same in branches 12 and 13 while the sinusoidal current phase reverses. The $L \cdot di/dt$ generated spikes are flowing as they would be without the bypass capacitors, that is in the same direction in branches 12 and 13. However, the sinusoidal current is flowing around the resonant circuit loop composed of branches 12, 19, 13, and 6. Thus its relative phase as picked up by the loop is reversed between branches 12 and 13.

7.3.3   What happens when the loop is placed so that three of its sides are parallel to branches 12, 19, and 13? Why?

Answer: The loop output due to $L \cdot di/dt$ spikes decrease in amplitude and that due to sinusoidal current increases in amplitude. This happens since the sinusoidal current adds into the three sides of the loop while the spikes induced from branches 12 and 13 tend to cancel each other because their like phase is subtracted around the pickup loop.

7.3.4   How does the loop output compare if branches 7, 8, and 14 are used in the last question?

Answer: Similar patterns are noted, that is the sinusoidal current is flowing in loops and the spike component of current is flowing in the same directions in branches 7 and 8. Although the resonant frequency may be slightly different the sinusoidal amplitude output from the pickup loop may be nearly what it was for branches 12, 13, and 19. This is because an array of tuned circuits all tuned to about the same frequency efficiently propagate noise to the furthest loop on the board. This effect could have impact on the design of circuit boards with a regularly repeated structure, such as on a memory board. One might consider changing the layout between sections or varying the size of the bypass capacitors on such a board so as to make adjacent parts of the circuit not resonant on the same frequency.

PROBLEMS AND DISCUSSION

7.1   What is a major limitation on using injected noise to measure the noise margin of a logic connection, such as an ALE lead?

Answer: While input threshold is relatively easy to measure this way, to measure the noise margin would require the injected noise to be coincident with the circuit noise so that the amount of injected noise would be the noise margin of the circuit. The problem is that this could require a long wait since the probability of a noise spike being coincident with an injected spike is **very** low. The amplitude of the injected noise would have to be raised very slowly, possibly over hours or even days, depending on the duty cycle of the noise being measured.

7.2   How can one determine which, if any, of several unsynchronized switching power supplies in a system are generating the noise that can be measured at a point in the system?

Answer: Display the noise on one trace on an oscilloscope and use this noise measurement as a source of a stable trigger for the oscilloscope as well. Connect a pickup loop to another vertical input of the scope. Hold this loop up to the power or ground lead of each supply to pick up some noise. Alternately, one could set the loop near the power supply circuitry if accessible. The power supply that is the source of the noise will be synchronized in time to the noise measurement while the other supplies will generate noise that will not be synchronized and will appear to precess across the screen or appear fuzzy. Only the power supply that is the source of the noise will appear stationary and clear on the scope.

8

Current Probes, Theory and Uses

INTRODUCTION

Current probes are one of the most useful measurement tools for making high frequency measurements on electronic circuits. They usually provide good isolation between the circuit and the measurement equipment, typically an oscilloscope or spectrum analyzer, because a direct connection to the circuit under test is not required. This isolation is probably the single most important characteristic of current probes. It allows one to measure electrical parameters of a circuit with a minimal effect on the circuit.

Many types of high frequency measurements, including voltage measurements, can be made with current probes. Since most high frequency circuits are relatively low impedance,[1] making a voltage measurement can be as simple as connecting a resistor between the nodes to be measured and using a current probe to measure the current through the resistor. Other types of measurements include the

---

1. As mentioned in Chapter 4, 10 pF of capacitance has only 160 ohms of reactance at 100 MHz. Only tuned resonant circuits can be high impedance at high frequencies.

degree of balance in a balanced circuit, sometimes referred to as common mode rejection, the *net* current in a bundle of wires, and phase measurements between two currents.

There are several types of current probes, ranging from devices resembling metal donuts to the simple pickup loops discussed in Chapter 7. They all involve some mechanism for directing the magnetic flux resulting from a current through a pickup device.

DC Coupled

In a popular implementation of a DC coupled current probe, the magnetic flux of the current to be measured flows in a core of ferromagnetic material that directs it to a Hall-effect solid state device. This device generates a voltage proportional to the magnetic field. Typically, the magnetic core has a sliding part that opens to accept the current carrying wire into the center of the core. One commercial device has response from DC to 50 MHz. Sometimes this type of probe can be susceptible to electrostatic fields which can cause it to indicate a DC current which is not real.

AC Coupled

AC coupled current probes are the most common and are available with up to 1 GHz frequency response. A typical current probe of this type uses a ferromagnetic core with the conductor passing through a hole in the center. The magnetic flux of the current is concentrated in the magnetic core around which is wound a pickup coil. The voltage induced in the pickup coil is proportional to the current flowing in the wire. Figure 8.1 shows a simplified diagram of such a device.

Current probes of this type are available in a number of physical implementations, but one of the most popular is one utilizing a split core with a hinge and latch so that the conductor to be measured for current does not have to be cut. Some current probes designed for permanent installation in equipment do not have this feature although their electrical operation is identical. Figure 8.2 shows a picture of several different types of AC coupled current probes including

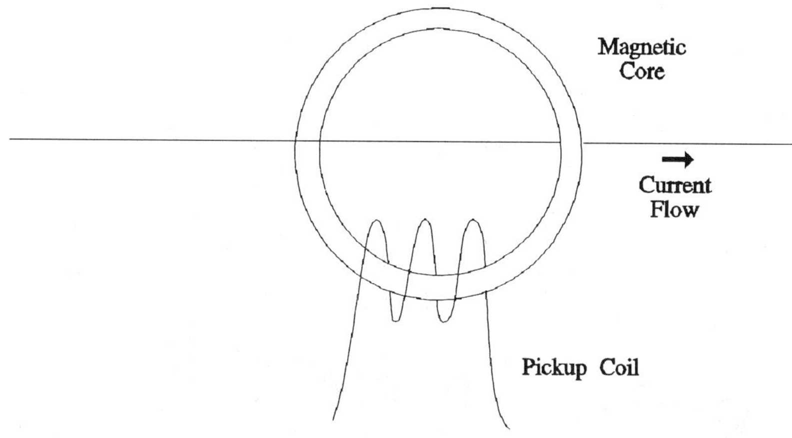

Figure 8.1 Simplified AC coupled current probe.

several variations on the split core and a surface current probe that
measure the current flowing in a conductive surface underneath the
probe. It is essentially one half of a split core probe.

Some of the probes shown in Figure 8.2 are referred to as being
of the "hinged donut" design. This type of current probe is very
popular for temporary troubleshooting.

The square magnetic pickup loop described in the last chapter is
a form of AC coupled current probe. In that case, there is no magnetic
core, the loop is oriented so that the magnetic flux of the current to
be measured passes through the center of the loop inducing a voltage
in it. The lack of a magnetic core has two effects. First, the magnetic
coupling from the current into the loop is variable depending on
insulation thickness and other factors which make absolute calibration of a pickup loop as a current probe difficult. Second, without a
magnetic core the inductance of the pickup loop is much smaller than
the pickup coil of magnetic core current probes. As will be seen
shortly, this causes the low frequency response as a current probe to
fall off at a much higher frequency than most magnetic core current
probes. Within these limitations though, the square pickup loop
operation is described by the same equations and principles of

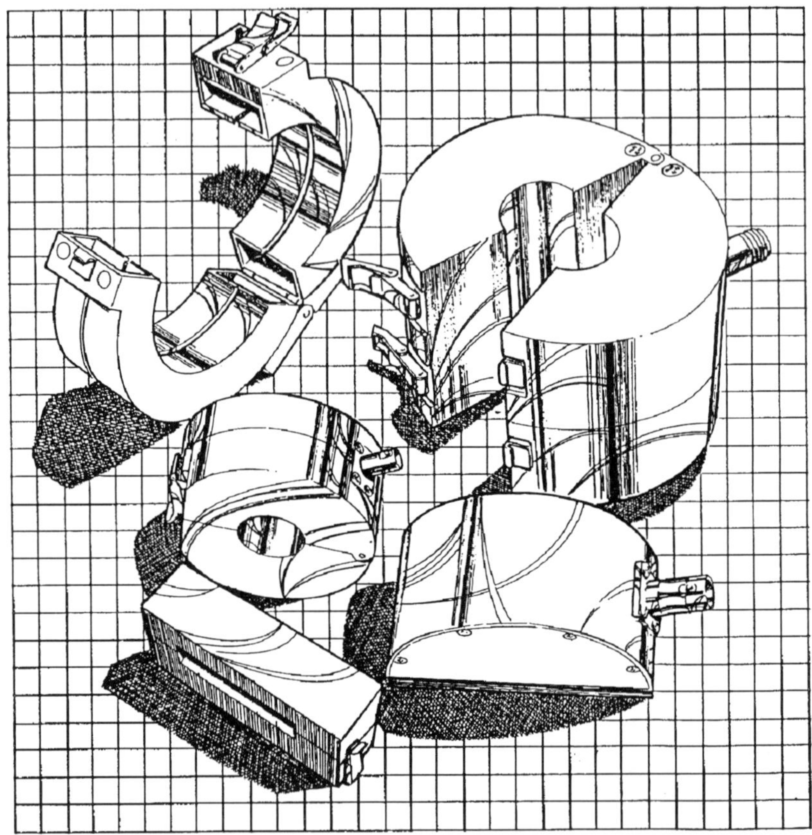

Figure 8.2 Examples of split core AC coupled current probes.[2]

operation. The rest of this chapter will be concerned with the operation and use of AC coupled current probes utilizing a magnetic core.

THEORY OF OPERATION

In order to make the best use of current probes it is necessary to develop a basic understanding of how they work. This section will

2. Courtesy of Fischer Custom Communications, 3121 W. 139th
    Street, Unit F, Hawthorne, CA 90250, Phone: 310-~~644-0728~~. 891-0635
    2905 W. Lomita Blvd, Torrance, CA 90505

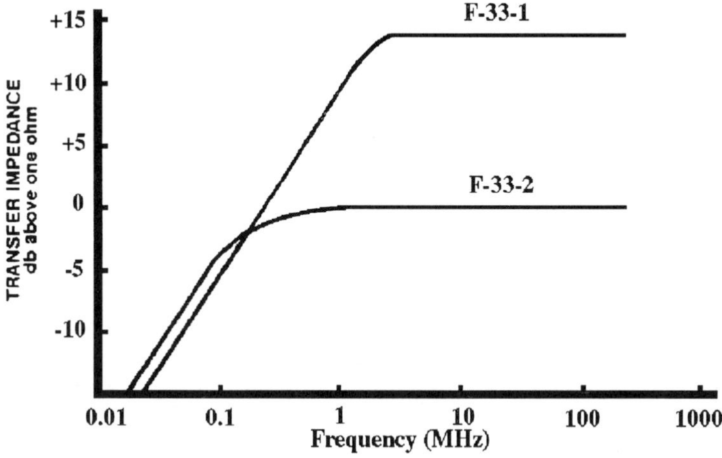

Figure 8.3 Transfer impedance vs. frequency for two probes.[3]

develop a basic theory of operation of AC coupled current probes, both with and without a magnetic core.

Transfer Impedance

Current probes are characterized by the parameter of "transfer impedance," the ratio of output voltage to current flowing through the current probe. Since it is the ratio of a voltage to a current, transfer impedance has units of ohms. In this case, the current is not caused by the voltage, as would be the case for the impedance of a network, since it occurs in a different circuit, but nonetheless this ratio can be defined, and as will be discussed in this chapter it is quite useful. Figure 8.3 shows a graph of transfer impedance versus frequency for two commercially available hinged donut current probes.

---

3.   Courtesy of Fischer Custom Communications, 3121 W. 139th Street, Unit F, Hawthorne, CA 90250, Phone: 310-644-0728.

Although transfer impedance is measured in ohms, it is often plotted on a log-log scale as in Figure 8.3. The ordinate in Figure 8.3 is labeled in dB above one ohm. This is calculated by:

$$dB\ ohms = 20\ log(|Z_t|) \qquad (8.1)$$

where $(|Z_t|)$ is the magnitude of the transfer impedance.

As plotted this way in Figure 8.3, the transfer impedance characteristic has a flat response for a range of frequencies where the probe is normally used. Below a corner frequency, F_c, the response falls off at 20 dB per decade, or 6 dB per octave, with decreasing frequency. Figure 8.4 shows a generalized plot of transfer impedance for a current probe.

In the figure, the output of the current probe for a constant current rises with frequency until it reaches the corner frequency, F_c, after which it remains constant with increasing frequency. The region below F_c, where the output rises with frequency, could be called the "voltage region" and the region above F_c could be called the "current region." This is the region where most current probes are used. The square magnetic pickup loop is used to measure noise voltages in the voltage region. These concepts will be made clear by looking at how a current probe works.

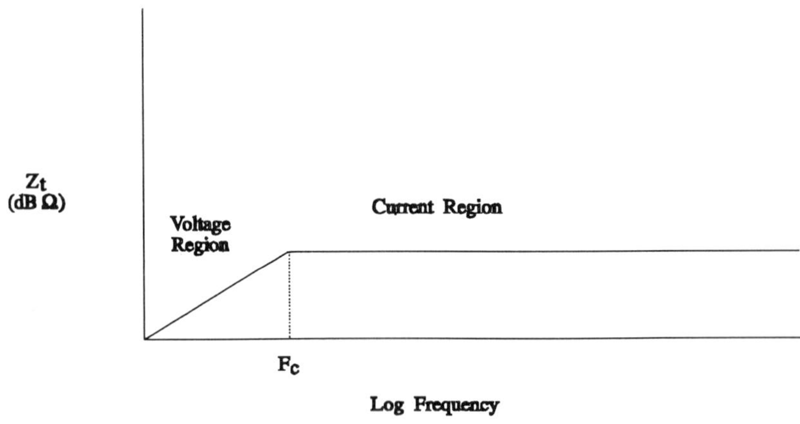

Figure 8.4 Transfer impedance vs. frequency—generalized.

Current Probe Equivalent Circuit

Any current probe, or clamp, as hinged donut probes are sometimes called, can be modeled as a transformer as shown in Figure 8.5. The primary winding of the transformer is the inductance of the wire passing through the probe, or near it for some designs such as surface current probes or magnetic pickup loops, including any effect of the probe core. The secondary winding is the pickup coil, the square magnetic loop itself or the coil around the probe core for a current probe with a magnetic core.

The open circuit, Thevenin equivalent, voltage induced into the secondary "winding" of Figure 8.5 is given by:

$$V_S(t) = M \cdot dI(t)/dt = M \cdot j \cdot \omega \cdot I(t) \tag{8.2}$$

where $\omega = 2 \cdot \pi \cdot f$ *and*

M is the mutual inductance between the pickup coil and
the conductor being measured.

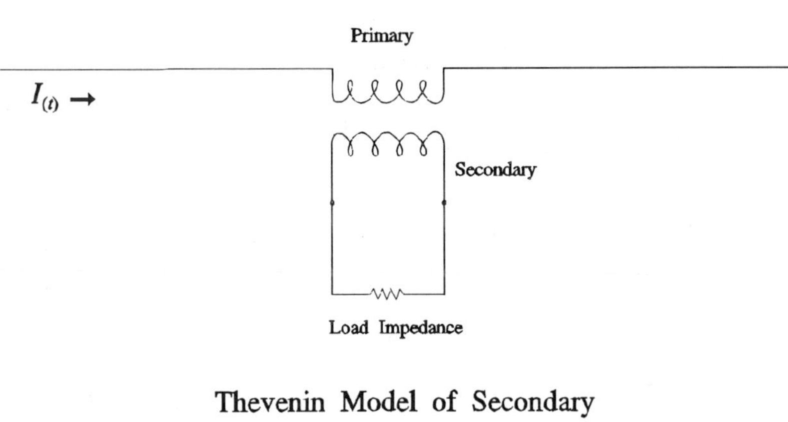

Thevenin Model of Secondary

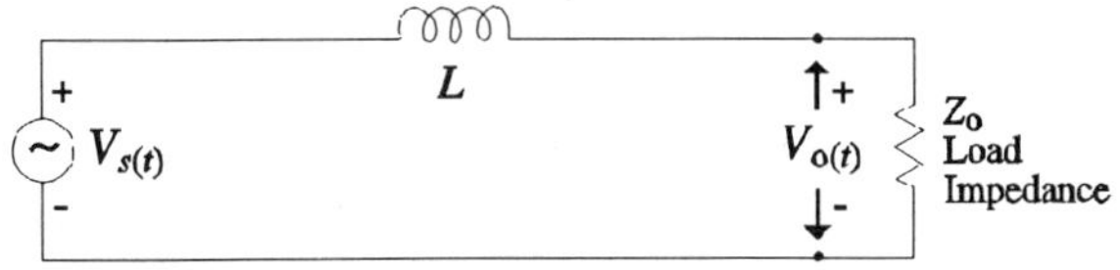

Figure 8.5   Simplified model of a current probe.

The equivalent Thevenin source impedance in series with the open circuit voltage is given by:

$$Z_S = j \cdot \omega \cdot L \qquad (8.3)$$

where L is the inductance of the secondary or pickup coil.

The transfer impedance is the combination of the open circuit voltage which rises with frequency at 20 dB per decade and the L/R low pass filter caused by the probe's self inductance and the resistive load, usually 50 ohms. The filter response is flat to the corner frequency, F_C, and then falls off at 20 dB per decade with increasing frequency. The result is the general transfer impedance graph given in Figure 8.4.

Equations 8.2 and 8.3 can be combined with the voltage divider equation to yield the result:

$$V_o(t) = M \cdot j \cdot \omega \cdot I(t) \cdot \frac{Z_0}{Z_0 + j\omega L} \qquad (8.4)$$

where V_0 is the voltage delivered by the probe to the load
Z_0 is the load impedance, usually 50 ohms.

Solving Equation 8.4 for the transfer impedance Z_t, $V_o(t)/I(t)$, and rearranging gives the transfer impedance as:

$$Z_t = \left[\frac{M \cdot j \cdot \omega}{1 + j\omega L/Z_0} \right] \qquad (8.5)$$

The corner frequency, F_C, occurs when $\omega L = Z_0$ and is given by:

$$F_c = \frac{Z_0}{2\pi L} \qquad (8.6)$$

Below F_C ($\omega L < Z_0$), in the voltage region, Equation 8.5 reduces to:

$$Z_t = j \cdot \omega \cdot M \qquad (8.7)$$

Notice that this is just the open circuit induced voltage into the "secondary" coil and is the equation used in Chapter 7 to estimate the $L \cdot di/dt$ voltage in a nearby conductor, hence the name "voltage region" for the low frequency part of the transfer impedance curve.

Equation 8.7 is just saying that at low frequency, the self inductance of the current probe pickup coil does not affect the probe output.

Above F_c ($\omega L > Z_o$), in the current region, Equation 8.5 reduces to:

$$Z_t = \frac{M}{L} \cdot Z_o \tag{8.8}$$

At high frequencies, the falling response of the low pass L/R filter is just cancelling the rising open circuit voltage for a constant current to give a flat response. Remembering the square magnetic pickup loop of Chapter 7 where M is always less than L, the transfer impedance for such a probe must be less than Z_o, usually 50 ohms. Many current probes have a transfer impedance of 1 to 5 ohms.

Additional Current Probe Characteristics

The question of what happens to the transfer impedance of a current probe if the number of turns in the pickup coil is increased has an interesting answer. Suppose one were to double the number of turns on the pickup coil of a current probe. How is the transfer impedance affected?

Since the self inductance of the coil is proportional to the square of the number of turns, L would increase by a factor of 4 in Equation

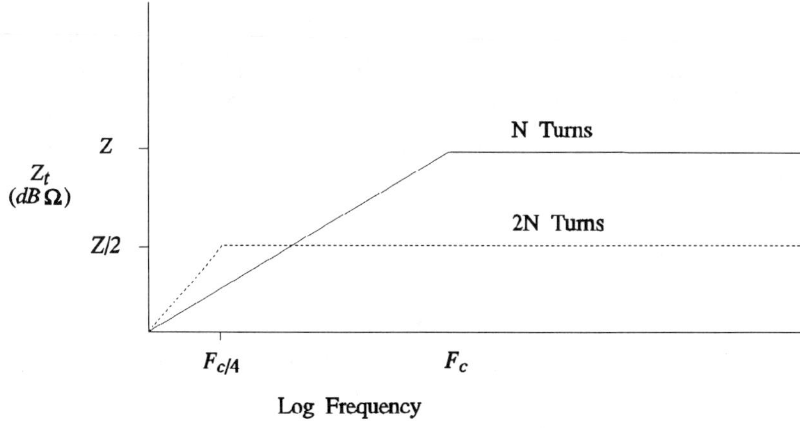

Figure 8.6 Transfer impedance with doubled pickup coil turns.

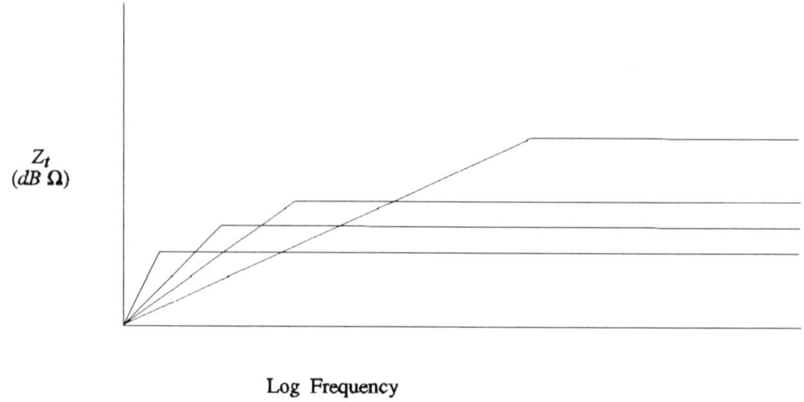

Log Frequency

Figure 8.7 Transfer impedance—family of curves.

8.8. However, M is proportional only to the number of turns as the number of secondary turns are increased, so doubling the number of pickup coil turns only doubles M in Equations 8.7 and 8.8. The result is that at low frequencies where the transfer impedance is only proportional to M, the transfer impedance will double, but at high frequencies where the transfer impedance is proportional to M/L, the transfer impedance will be cut in half. The situation is pictured graphically in Figure 8.6. Chapter 7 contains a graphical treatment of this subject as applied to the square pickup loop.

Below F_c in Figure 8.6 the slope of the transfer impedance curve is doubled by doubling the number of turns in the pickup coil, however F_c itself is lowered by a factor of 4 at the same time since it is inversely proportional to L. So, the sloped portion of the curve intersects the flat portion at one half the frequency and one half of the magnitude of transfer impedance as well. The effect of doubling the number of pickup coil turns is then to produce a probe that is one half as sensitive, but with flat frequency response to a frequency that is lower by a factor of 2. By adjusting the number of turns, one could create a family of curves shown in Figure 8.7.

The effect of load impedance, Z_o in the above equations, on a current probe is also interesting. This effect will be demonstrated and explained in Experiment 8.1 at the end of this chapter.

Applying the Transformer Equation to Current Probes

For current probes employing a magnetic core, the follow relationship for the mutual inductance in magnetic core transformers, M, holds:

$$M = k \cdot (L_1 \cdot L_2)^{1/2} \tag{8.9}$$

where: k is the coefficient of coupling,
$\qquad L_1$ is the inductance of the primary, and
$\qquad L_2$ is the inductance of the secondary.
As applied to current probes, L_2 is the inductance of the pickup coil in the probe, called just L up to this point, and L_1 is the inductance of the wire passing through the probe, usually somewhat increased by the probe's presence.

Substituting M from Equation 8.9 into Equation 8.8 and using the terms L_1 and L_2 from Equation 8.9 yields:

$$|Z_t| = \frac{M}{L} \cdot Z_0 = \frac{k \cdot (L_1 \cdot L_2)^{1/2}}{L_2} \cdot Z_0 = k \cdot \frac{L_1^{1/2}}{L_2^{1/2}} \cdot Z_0. \tag{8.10}$$

Equation 8.10 leads to an interesting result which becomes obvious when one realizes that the turns ratio of a transformer is the square root of the inductances of the windings. Since L_1 represents the wire passing through the center of the current probe and usually represents *one* turn, then Equation 8.10 can be written as:

$$Z_t = k \cdot \frac{Z_0}{number\ of\ turns\ on\ L_2} \tag{8.11}$$

where Z_t is the transfer impedance above F_c.

Using Equation 8.11, it is possible to estimate the number of turns needed on the pickup coil for a desired transfer impedance. Knowing the turns ratio, it is possible to estimate the series voltage induced in the wire passing through the current probe. This is useful in determining the effect of the current probe on the circuit under test.

Using Equations 8.6, 8.8, and 8.9, one can solve for the probe parameters M, L_1, and L_2. These become:

$$L_2 = \frac{Z_0}{2\pi F_c} \tag{8.12}$$

where: L_2 is the inductance of the pickup coil in the probe.

$$M = L_2 \cdot \frac{Z_t}{Z_0} = \frac{Z_t}{2\pi F_c} \qquad (8.13)$$

where: M is the mutual inductance between the pickup coil and the measured conductor,

Z_t is the transfer impedance above F_c.

$$L_1 = \left[\frac{M}{k}\right]^2 \cdot \frac{1}{L_2} \qquad (8.14)$$

where: L_1 is the inductance of the measured conductor that passes through the probe.

The parameter of most interest to many users of a current probe, of the three given above, is L_1, for that is the inductance that the probe injects into the circuit being measured if the probe is not loaded by Z_0 on its output and does not have an internal load either. Some probes have a built in termination impedance that in parallel with the external load forms the load impedance on the current probe, Z_0 in all of the previous equations. These probes reflect that termination impedance back across L_1 and thus reduce the effect of an unconnected probe on a circuit.

Be careful. L_1 can be substantial for some current probe designs. It is best to keep current probes terminated in their recommended load impedance any time they are installed on a circuit even if not being used all of the time.

Calculations for Commercial Probes

Table 8.1 shows advertised and calculated numbers for three commercial current probes. The assumption is that the coefficient of coupling, k, for each probe is unity. If it is less than that, L_1 becomes larger.

Probe 3 has a bandwidth from about 3 MHz to about 300 MHz. Its pickup coil around the magnetic core may have as many as 10 turns (if $k = 1$, less for smaller k). To keep the frequency response flat over such a wide range of frequency the design of the pickup coil, core, and other parts of the probe is very critical. Small parasitics must be well controlled to achieve the desired result.

Table 8.1. Current probe examples.

	Probe 1	Probe 2	Probe 3
Advertised			
$Z_t\,(\Omega)$	5	1	5
$Z_o\,(\Omega)$	25^4	25^4	50
F_c	25 kHz	1.2 kHz	3 MHz
Calculated			
L_2	159 µH	3.32 mH	2.65 µH
M	31.8 µH	133 µH	265 nH
L_1	6.36 µH	5.33 µH	26.5 nH
Turns on L_2[5]	5	25	10

USES FOR CURRENT PROBES

There are many uses for current probes. Chapter 9 describes their use for measuring high peak currents with amplitudes on the order of amperes caused by pulsed electromagnetic interference such as electrostatic discharge. Current probes are often used to measure very small currents, on the order of microamperes, as well. One such use is to measure common mode currents on cables to predict radio frequency emissions as related to the legal requirements of FCC Part 15 rules for computing equipment.

Common Mode Current Measurements

The radiation from a cable as measured as the magnitude of electric field strength some distance from the cable caused by a common

---

4. Has built-in 50 ohm load so total load on the current probe with a 50 ohm cable is 25 ohms.
5. Assuming k is unity.

mode current is proportional to the cable length, up to one quarter wavelength, the frequency of the common mode current, and the amount of current as shown in Equation 8.15.

$$E \propto f \cdot L \cdot I \qquad (8.15)$$

where: E is the electric field strength,

f is the frequency of the common mode current,

L is the length of the wire ($< 1/4$ wavelength), and

I is the magnitude of the common mode current.

For a cable 1 meter in length at 75 MHz, a common mode current above 15 microamps will possibly cause a failure of FCC Part 15 Class A rules for the emissions allowed from computing equipment[6]. Even though all of the assumptions that were made may not apply to every situation, practical experience has shown the 15 microamp limit to be useful. By measuring the common mode current on equipment cables before the equipment is sent off for official emissions testing, several cycles of the fail-retest-fail sequence can be avoided.

Isolated Voltage Measurements

Current probes can be used to make isolated voltage measurements as well. The idea is to connect the two nodes between which the voltage is to be measured, such as two points on circuit ground of a printed wiring board, with a known impedance. Install the current probe around the leads connecting the known impedance to the circuit. Once the current through the impedance is known, the voltage between the two nodes can be calculated.

To work properly, the known impedance, usually a resistor, should be large in value compared to the series inductive reactance of the wire used at the highest frequency of interest. The known impedance should also be low enough in value to allow enough current to flow to make a reasonable current measurement. For most low impedance high

---

6.   More information on this topic is available in *Noise Reduction Techniques in Electronic Systems,* Second Edition,H. W. Ott, John Wiley & Sons, Inc., Chapter 11.

frequency circuits, a compromise value for the known impedance can be found to meet both of the above requirements.

Relative Phase Measurements

Using two matched current probes, the phase between two currents can be measured. This information can be quite useful. Such was the case using two pickup loops to find a noise source in Chapter 7. In addition, knowing the phase between two currents can be used for the debugging of equipment or design verification. The next chapter contains more detailed information on how to make this type of measurement and its uses.

LIMITATIONS

Current probes have limitations that must be taken into consideration in order to get accurate measurement results. Several of these limitations are discussed in this section.

Capacitive Coupling to Circuit Under Test

Almost all current probes have enough capacitance, as much as a few pF, from the probe to the circuit under test to cause potential problems at the higher frequencies, especially at 1 GHz and above. One commercially available probe has a probe to circuit capacitance of one picofarad and the probe's bandwidth extends to 1 GHz. At a frequency of 1 GHz, one picofarad of capacitance has a reactance of only 160 ohms. This level of impedance is on the same order as the circuits likely to be measured. Loading down a 100 ohm circuit with 160 ohms of parasitic capacitive reactance is likely to lead to measurement errors as well as to affect circuit operation. It is possible that a substantial fraction of the current measured may actually arise through this parasitic capacitance. The best defense against problems of this type is to use current probes at points in the circuit with low voltages with respect to ground (the body of the probe). For example:

measure the current on the ground side of a load instead of the signal side if possible.

Adding Series Impedance to Circuit Under Test

Be sure to take into account any series impedance added to the circuit under test by the current probe. Sometimes this number is stated in the probe specifications, especially for probes with built-in terminations and k near unity. Some probe designs do not add a significant series impedance, such as Probe 3 in Table 8.1. That probe can only add a few tens of nanohenries, equivalent to a few inches of wire, even if left unterminated. To estimate series impedance, if not specified, calculate L_1 from Equations 8.12, 8.13, and 8.14 assuming that the probe does not have an internal termination, use one half as a reasonable estimate for k.

Be sure to terminate the probe to minimize added series impedance. The added series impedance caused by the probe can be reduced to as low as as the termination impedance divided by the square of the turns ratio, for $k = 1$. The turns ratio can be estimated from Equation 8.11. Usually the impedance reflected into the circuit under test by the probe termination is low enough not to be a problem.

Effect of Transfer Impedance on Equipment Operation

Transfer impedance is a measure of how closely a current probe is coupled to the circuit under test. Equation 8.8, repeated here, shows that transfer impedance is proportional to M, the mutual inductance between the probe pickup coil and the circuit under test.

$$Z_t = \frac{M}{L} \cdot Z_o \qquad (8.8)$$

A current probe with only 1 ohm of transfer impedance delivers a signal to its 50 ohm load that is only one-fifth of what a probe of 5 ohms transfer impedance will. Assuming that the lower transfer impedance was not achieved by simply loading the output of the current probe, additional energy came from the circuit under test to generate the higher output of the 5 ohm probe. The greater the mutual coupling between the circuit under test and the current probe, the greater is the possibility of affecting the circuit operation.

Spurious Response

Many current probes have a significant response to electric fields. A response that can also cause error in a current measurement. Hopefully the spurious electric field response is at least 20 dB below the current response for a particular measurement. This is most likely to be a problem when the measurement is made on a wire that has a high electric field associated with it. Situations in which current probes are likely to be used under such conditions are covered in the next chapter. A few methods to measure the electric field response of a current probe are also discussed there. The best defense is to keep the potential difference between the current probe and the circuit under test to a minimum by putting the probe on the ground side of a circuit, when possible.

Magnetic Core Saturation

This problem is not likely for most uses of current probes. Currents high enough to saturate probes can be caused either directly by the high frequency currents to be measured, or by lower frequency currents which are present and which are not part of the measurement. High currents are generated, for example, by pulsed electromagnetic interference, discussed in the next chapter, and measurements involving the AC power line. When using current probes in these environments, check the probe specifications to make sure that the probe is capable of operation in the presence of currents that are likely to be present. Probes with small cores are the most suspect. A current rating of 50 to 100 amperes peak is not unusual for a hinged donut design of a few inches in diameter. This is adequate even for most pulsed electromagnetic interference measurements.

SUMMARY

Current probes are extremely useful tools for making high frequency measurements. A main advantage is the lack of any direct connection to the equipment under test and the resultant minimal impact on its operation. This chapter concentrated on AC coupled current probes with magnetic cores.

The theory of operation of AC coupled current probes was developed and several design equations for the probe derived. These equations are useful from the current probe user's point of view as they provide insight into how the probes work, illuminate possible sources of error in the measurement and some limitations of current probes, and ways of improving probe performance. Calculations using the design equations for a few commercially available probes were made to show the relative magnitudes of quantities that affect circuit operation such as the series impedance developed in the circuit under test by the probes.

Current probes are useful for several types of measurements including:

- Common mode current measurements to estimate electromagnetic radiation from cables (currents in μa, frequencies from the tens to hundreds of MHz).
- Voltage measurements where the measuring equipment is isolated electrically from the circuit under test (currents in ma, frequencies from the tens to hundreds of MHz).
- Relative phase measurements between currents.
- Pulsed electromagnetic interference (peak currents in the tens of amperes with risetimes to less than one nanosecond, discussed in more detail in Chapter 9).

Current probe limitations include:

- Capacitive coupling between the probe and the circuit under test.
- The addition of series impedance into the circuit under test.
- The effect of high transfer impedance on the circuit operation.
- Spurious response, especially to electric fields.
- Magnetic core saturation.

For most measurements, these factors are not a problem. However by understanding how current probes work, one can minimize their affect in difficult measurement situations such as near electrostatic discharge.

LABORATORY DEMONSTRATIONS

EXPERIMENTS

Equipment needed:

- 100 MHz or higher bandwidth scope (50 ohm input is desirable).
- 5-50 MHz square wave oscillator with rise/fall times less than 5 ns such as described in Chapter 3, Figure 3.8.
- Two 5-200 MHz bandwidth hinged donut current probes with a 5 ohm transfer impedance such as Fischer Custom Communications F33-1 (probes that have matched response are desirable).
- Five 50 ohm feed-through coaxial cable terminations.

Experiment 8.1: Current Probe Low Frequency Response

1. Connect the output of the oscillator through a 50 ohm resistor and 6 inches of wire back to its signal ground. Place a current probe around the wire and connect to the oscilloscope and terminate the probe cable in 50 ohms.
2. Set the oscillator for 5 MHz operation. The oscillator will produce several tens of milliamperes of square wave current in the wire so the probe will have 5 times that much output. Set the vertical scale on the scope to about 100 mv per division to get a waveform a few divisions in height and the horizontal scale to about 200 nanoseconds per division. Note that the waveform displayed will not be a good square wave but have some "droop."
3. Put four 50 ohm feed-through terminations on the current probe output jack and connect the scope cable to the last termination on the probe. Now the load on the probe is 10 ohms instead of

50 ohms as it was in step 2 above. Adjust the vertical sensitivity as necessary to maintain a waveform of a few divisions height. Note how the waveform changed from step 2.

Questions:

8.1.1   What causes the "droop" in the squarewave as displayed on the scope?

Answer: Since the square wave frequency is near the lower end of the current probe frequency response, the fundamental of the squarewave is somewhat attenuated and so the top and bottom of the squarewave (which is supposed to be flat) drops or "droops" some fraction of the distance to the baseline.

8.1.2   How does the response in step 3 compare to step 2 and why?

Answer: The waveform will be about one-fifth the height and the "droop" will almost disappear. By adding the extra load to the output of the probe to make it 10 ohms instead of 50 ohms, the L/R corner frequency of the probe response is lowered by the same factor of 5. The corner frequency is now well below the squarewave fundamental frequency and the squarewave of current looks "square" on the scope. The amplitude of the squarewave is also reduced by the same factor of 5 because the lower corner frequency flattens the rising portion of the probe response at one-fifth of the original frequency. This results in the probe having a 1 ohm transfer impedance. This effect is shown graphically in Figure Ex8.1.

Lowering the load impedance on the probe is an excellent way to improve low frequency response at the expense of transfer impedance. For many uses, the signals are large enough that a lower transfer impedance is acceptable. If the load impedance is made low enough to become comparable to the winding resistance of the probe pickup coil additional loss will result and the transfer impedance will be lower than expected. When lowering the load impedance of a current probe, always check the probe's response on a known signal.

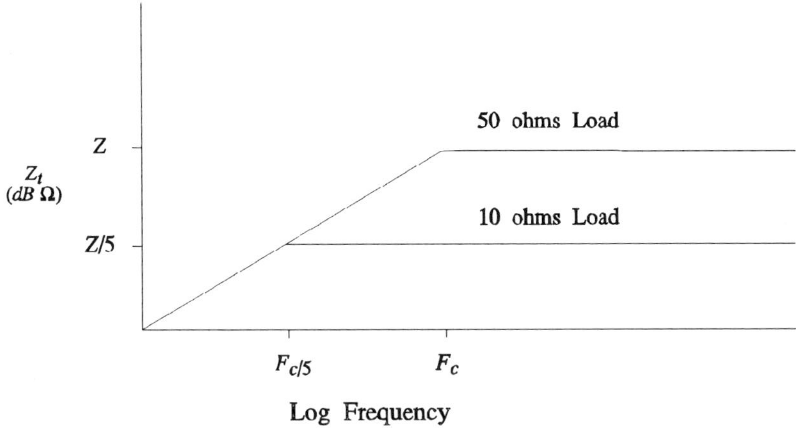

Figure for Example 8.1   Load impedance effect on probe transfer impedance.

Experiment 8.2: Common Mode Current Measurment

1. Connect a 6 foot length of 50 ohm coaxial cable to the square-wave generator and install a 50 ohm feed-through termination on the end of the cable. Set the generator to operate between 30 and 50 MHz.
2. Install a current probe on the coaxial cable near its middle. Connect the probe cable to a 50 ohm termination at the scope and set the scope vertical scale at about 10 mv per division and the horizontal scale at about 50 nanoseconds per division.
3. Vary the frequency of the generator between 30 and about 50 MHz and note the scope display.
4. Connect a 6 foot length of wire to the center pin of the 50 ohm feed-through termination at the end of the coaxial cable as shown in Figure Ex8.2.
5. Vary the frequency of the oscillator and note the scope display.
6. Move the current probe up near the termination but still over the coax and install another one over the 6 foot wire near the termination as well. Be sure to connect the second current probe to the scope with the same type and length of coaxial cable so that the time delays from each probe to the scope are equal.

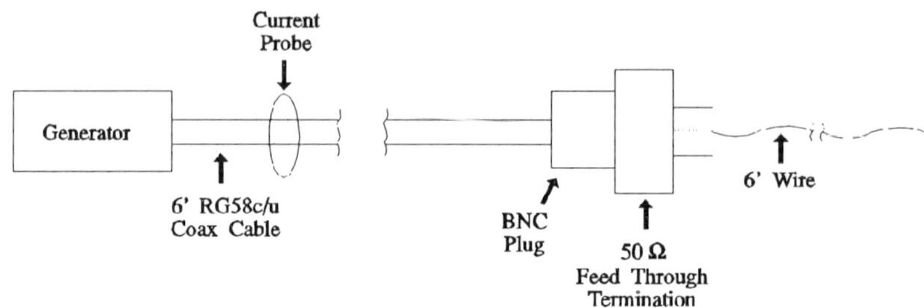

Figure for Example 8.2   Experimental setup for Experiment 8.2.

Display both waveforms on the scope and vary the frequency from 30 to 50 MHz, as before noting the scope response.

7. Move the two current probes together over the coaxial cable near the termination and junction of the wire. Subtract the two signals in the scope using A-B and note the result as the probes are separated along the coaxial cable.

8. Remove the 6 foot wire leaving just the coaxial cable and its termination. Install a 6 inch pigtail on the coax shield at the oscillator end of the coaxial cable. Vary the frequency of the generator and note the output from a single current probe located near the middle of the coaxial cable.

Questions:

8.2.1 How do the scope displays in steps 3 and 5 compare as the frequency of the oscillator is varied?

Answer: In step 3, there will be no net current on the coax cable and no output from the current probe because all of the signal current flowing on the center conductor of the coax passes through the load resistor and returns on (the inside of) the shield. However, in step 5, the addition of the 6 foot wire to the center conductor beyond the load resistance changes the currents in the coax cable significantly.

As the frequency is varied, the current probe will indicate peaking of current on the order of tens of ma peak at certain frequencies. At these frequencies, the wire has become resonant and the impedance looking into the end of the wire appears comparable to 50 ohms. The result is that some of the current flowing on the coax center conductor flows into the wire and is radiated. An equal and opposite current flows on the outside of the shield. Since the total shield current no longer equals the center conductor current in the coax, there is a net current and therefore an output from the current probe.

8.2.2   How do the outputs of the two current probes compare in step 6?

Answer: The current that flows in the wire is just the amount that the center conductor and total shield currents differ in the coax so that the two probes will show the same output as long as they are within a few inches of each other.

8.2.3   What happens to the scope display in step 7 as the probes are moved apart? Why?

Answer: With the two probes next to each other they should read equal currents (if the probes are matched) so when the two signals are subtracted, the net result is almost zero. As the probes are separated on the cable both the magnitude and phase read by each probe will differ from each other and the difference between the two signals will increase and be visible on the scope. The further the probes are separated, up to a point, the larger the difference signal becomes.

This is caused by the fact that at these frequencies, a few feet is a significant portion of a wavelength and so a current will undergo substantial phase shift in that distance. The common mode current flowing on the cable also results in standing waves which change the current amplitude as well over distances of a few feet.

8.2.4   What does the output of the probe look like in step 8 and why?

Answer: The installation of the pigtail at the generator end of the coax cable causes many effects, one of them being a phase shift in the current on the shield with respect to the center conductor. This phase shift causes the center conductor and shield currents not to cancel well. The net result is that, at some frequencies between 30 and 50 MHz. from the square-wave generator, there will be a net common mode current on the coax cable of a few ma or so.

PROBLEMS AND DISCUSSION

8.1   How can one estimate the series voltage induced in the circuit under test by the presence of a current probe from the current probe output?

Answer: Below F_c, where the effect of the self inductance of the probe pickup coil is low, just divide the output of the current probe by the turns ratio (the number of turns on the pickup coil) and then multiply that number by $\frac{1}{k}$ to get the voltage induced in the wire being measured.

For example assume a probe with a 10 MHz F_c, a 5 ohm transfer impedance in the flat region, a load impedance of 50 ohms, and that has a 2 volt output at 1 MHz. Assume a value for k of one-half, if it is not known, to calculate the number of turns on the probe pickup coil using Equation 8.11.

$$Z_t = k \cdot \frac{Z_0}{number\ of\ turns\ on\ L_2} \tag{8.11}$$

where Z_t is the transfer impedance above F_c.

From Equation 8.11, L_2 has 5 turns. Thus, the voltage induced in the circuit being measured would be $\frac{2}{5}$ Volt times $\frac{1}{k}$ or 800 mv.

Above F_c, one must multiply the above calculation by the frequency of the measurement divided by F_c to take into account the low pass filter function of the load impedance and the inductance of the probe pickup coil. So a 2 volt output at 20 MHz for the above probe would require 1.6 volts induced in series with the circuit under test.

9

Measurements of Pulsed EMI Effects on Electronic Circuits

INTRODUCTION

Measurements of high frequency voltages and currents in an electronic system can be used to determine the mechanisms responsible for system malfunction caused by a susceptibility to the electromagnetic environment. The most severe form of electromagnetic interference to digital electronic circuits is pulsed Electromagnetic Interference, pulsed EMI. Two common forms of pulsed EMI are Electrostatic Discharge, ESD, and Electrical Fast Transient, EFT. Both of these phenomena are related to electrical sparks, especially those occurring between conductors. There are internationally accepted tests for equipment susceptibility to these forms of pulsed EMI[1]

Unfortunately the same characteristics of pulsed EMI that make it so much of a problem for electronic systems, especially high amplitude voltages and currents with fast risetimes, also cause

---

1.   ESD equipment level testing is covered in IEC (International Electrotechnical Commission) standard 801-2 and EFT equipment level testing is described in IEC 801-4. Both of these documents are available from the American National Standards Institute (ANSI).

difficulty in making measurements of what is happening inside the system. This chapter will cover techniques for measuring voltages and currents both internal to and external to electronic systems for the purpose of determining the mechanisms of system susceptibility. Of course, the same techniques are useful for other high frequency measurements that may experience severe interference as well. Techniques for preventing interference to measurements as well as minimizing measurement effects on the system under test are presented. The interpretation of measurement results will also be discussed.

This chapter will first cover background information followed by typical measurement pitfalls. Realistic options for measurement of internal and external system voltages and currents in the presence of pulsed EMI will then be covered in detail. Finally, a discussion of the advantages of the measurement approach to equipment pulsed EMI susceptibility versus the more common trial and error method will be presented. As with many other chapters in this book, this chapter will end with laboratory demonstrations and questions.

TECHNICAL BACKGROUND

Voltages and currents on conductors induced by pulsed EMI events can be *very* large. These voltages and currents result from coupling of the sometimes extreme electric and magnetic fields from the pulsed EMI event to the equipment under test. The magnitude of the problem can be understood in light of the following two examples.

Inductive Coupling

Magnetic field coupling can result in large voltages induced into circuits by a pulsed EMI event. Consider magnetic flux, F, defined as:

$$\Phi = L \cdot I \qquad (9.1)$$

where: L is the inductance of a closed path, and

I is the total current flowing around the path.

By differentiating Equation 9.1, the following equation results:

$$d\Phi/dt = E = L \cdot dI/dt \qquad (9.2)$$

where: E is the voltage induced around the loop.

EXAMPLE

9.1   For: $L = 20$ nh/inch (a typical value, see Chapter 2)
              $dI = 1$ ampere
              $dt = 1$ nanosecond
       then: $E = 20$ volts/inch.

In Example 9.1, a current change of one ampere in one nanosecond was assumed to flow in 1 inch of wire. These are very moderate numbers. ESD currents flowing on cables, with currents of many amperes and with risetimes much smaller that one nanosecond, can easily generate hundreds of volts per inch along the cable's inductance. But, even with the assumed numbers, 20 volts per inch results. This is certainly enough to exceed the noise margin of digital logic and foil attempts to measure logic signals inside the equipment.

Expressing currents as $L \cdot di/dt$ in units of volts per inch is a useful way of estimating the interference potential of a noise current, such as that from a pulsed EMI event, to electronic circuits. For example, a current that results in 10 millivolts per inch on a printed wiring board is unlikely to result in malfunction of digital logic. There are simply not enough inches to generate a voltage to be comparable to digital logic noise margins. But, a current producing 200 millivolts per inch may cause interference to digital logic and a current causing 20 volts per inch almost certainly will interfere with the operation of digital logic.

Capacitive Coupling

Electric field coupling can result in large currents induced into circuits by a pulsed EMI event. Consider the following:

$$Q = C \cdot V \qquad (9.3)$$

where: Q is electric charge stored on two conductors,
C is the capacitance between the conductors, and
V is the voltage between the conductors.
By differentiating Equation 9.3 the following equation results:

$$dQ/dt = I = C \cdot dV/dt \qquad (9.4)$$

where I is the current flowing in the "plates" of the capacitor.

EXAMPLE

9.2   For: $C = 1$ picofarad
$dV = 1000$ volts
$dt = 1$ nanosecond
then: $I = 1$ ampere.

As in Example 9.1, these values are moderate, the current can be much more than one ampere though a picofarad.

Examples 9.1 and 9.2 serve to point out just how difficult measuring the effect of a pulsed EMI event can be. Small parasitic elements of the measurement setup, such as the inductance on an inch of wire or 1 picofarad of capacitance, can result in large parasitic voltages and currents. These parasitic voltages and currents are not only a source of error in measurements, but sometimes can cause the process of making a measurement to adversely affect the operation of the equipment whose response is to be measured. Therefore, special techniques must be used to accurately measure system voltages and currents in response to pulsed EMI. That is not to say expensive equipment is necessary, only that good technical judgement be used with the measurements, including performing appropriate null experiments. Often, easily built or relatively inexpensive commercially available measurement apparatus works very well.

The Skin Effect

In order to understand the measurement concepts presented in this chapter it is necessary to first understand the skin effect, the process whereby current flows near the surface of a conductor at high frequencies. This effect can be illustrated by the example shown in Figure 9.1.

In this example, a 10 MHz signal source is connected by wires to the center of a 1 foot square copper plate that is $\frac{1}{8}$ inch thick. The plate completes the path of the signal current between the ends of the two wires. Figure 9.1 shows a top and side view of the circuit. The current must either pass through the $\frac{1}{8}$ inch thickness of the copper plate or flow on the outside surface of the plate and around the edges to complete the circuit, a distance about 100 times longer than going through the plate. Which path will the current take?

The 10 MHz current will not flow through the plate. The skin effect causes it to flow near the surface of the plate and around the edges. The concept of skin effect is important when considering the design of system cabinets and the operation of shielded cables.

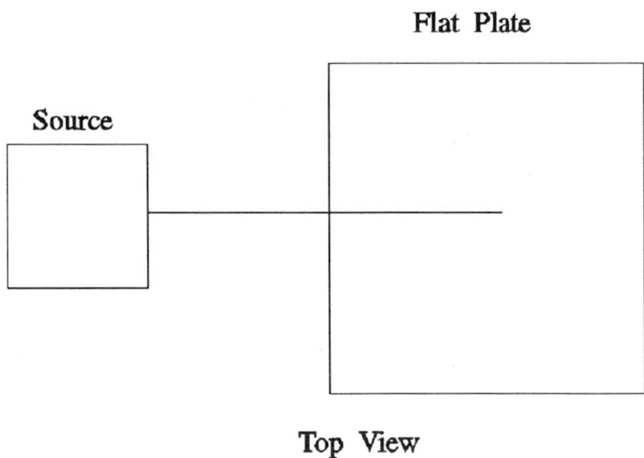

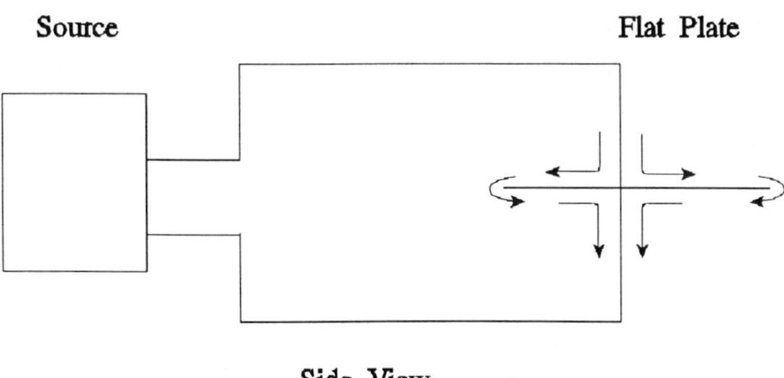

Figure 9.1 Flat plate example (of high frequency current flow).

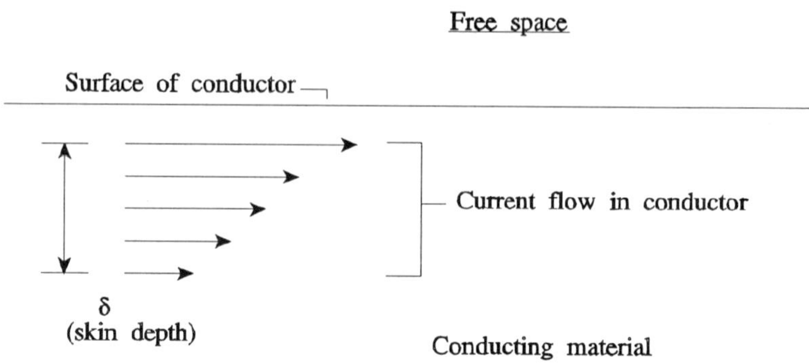

Figure 9.2 High frequency current flow in a conductor.

Current flow near the surface of a conductor at high frequencies is illustrated in Figure 9.2. As a high frequency current flows in a conductor, the current density falls off rapidly with depth into the conductor as a result of the skin effect. The skin depth of the current is defined as the depth, d in Figure 9.2, into the conductor at which the current density has fallen to $\frac{1}{e}$ or 37 percent of its surface value.

The skin effect raises the resistance of a conductor with increasing frequency since the current is not carried uniformly in the conducting material. For example, between DC and 5 MHz, the resistance of a 6 inch section of 24 gauge copper wire at room temperature will increase from 0.024 ohm to 0.055 ohm, slightly more than a factor of two.

For the purposes of this chapter, crowding of current near the surface of a conductor is more important for controlling where currents induced by pulsed EMI will flow rather than for the apparent increase in the resistance of a conductor. In fact, as discussed in Chapters 2, 4, and 5, the resistance of a conductor is generally negligible compared to its inductive reactance at frequencies above a few tens of kHz. Skin depth, d, is given by Equation 9.5 as[2]:

------

2.   For more information on skin effect see *Noise Reduction Techniques in Electronic Systems,* Second Edition, H. W. Ott, published by John Wiley & Sons, Inc., pp. 162-169.

$$\delta = \left[\frac{2}{2\pi f \mu \sigma} \right]^{\frac{1}{2}} \qquad \delta = \frac{6.6}{(f\,\mu_r \sigma_r)^{\frac{1}{2}}}\ cm \qquad (9.5)$$

General Form   *Practical Form*

where: σ_r is the conductivity relative to Cu,

μ_r is the relative permeability,

f is the frequency,

δ is the skin depth.

For copper, Equation 9.5 reduces to:

$$\delta \approx \frac{6.6}{\sqrt{f}}\ cm \qquad (9.6)$$

At about 10 MHz, the skin depth of copper is about 0.002 cm. Even lead which is less than one tenth as conductive as copper, and thus has a skin depth about 3.5 times that of copper, still has a skin depth at 10 MHz of only 0.007 cm.

For most metals and frequencies above a few tens of kHz., the skin depth is much less than the thickness of metal needed to provide reasonable physical strength. This is especially true for equipment enclosures. Possible exceptions to this general rule are conductive films such as metal film covered plastic and some conductive paints. For these, it is possible for the film to be thin enough or not conductive enough to provide adequate thickness, a couple of skin depths, for high frequency currents to flow freely.

The skin effect is very important to measurement techniques in an environment with an extreme level of high frequency interference, such as that generated by ESD or other forms of pulsed EMI. Due to skin effect, high frequency currents flowing on the inside of the shield of a shielded cable or the inside of a metal cabinet are distinct from high frequency currents flowing on the outside of the shield or cabinet. Skin effect also provides the ability to dump high frequency noise currents flowing on the outside of a cable shield, such as ESD induced noise, to a chassis in order to keep the noise currents out of equipment circuitry.

MEASUREMENT PITFALLS

Because of the large voltages and currents developed by small parasitic elements, measurements made near pulsed EMI are prone to error. In addition, the measurement can itself cause the equipment under test to malfunction by conducting noise into the equipment unless adequate precautions are taken. In this section, pitfalls of various measurement techniques as related to pulsed EMI troubleshooting will be covered.

One way to ensure accurate measurements is through null experiments, first discussed in Chapter 4. A null experiment is one that yields a known, often zero, result (hence the name *null experiment*). Its purpose is to determine the amount of error in a measurement caused by problems in the measurement equipment or setup, such as susceptibility to interference generated by pulsed EMI. A good null experiment validates the data taken in a given experiment. Null experiments are *very* important in pulsed EMI measurements on electronic systems because of the high potential for the interference to affect the accuracy of the measurement. Especially since much of the commonly available measurement methods and equipment were never designed to be used around pulsed EMI and the strong electric and magnetic fields that it generates. *All* measurements must be verified where pulsed EMI is involved.

Scope Probes

Scope probes can both be susceptible to pulsed EMI generated interference and significantly affect the equipment under test response to pulsed EMI. As shown in Chapter 4, the probe ground lead can be a significant source of error in measurements involving scope probes. In addition, since a probe requires a direct connection to the equipment, there is the possibility that conduction of pulsed EMI interference into the equipment under test will occur. In general, scope probes should not be used in a pulsed EMI environment to measure transient (one shot) signals.

Figure 9.3 illustrates the problem of scope probe susceptibility. A scope probe shorted by its own ground lead forms not only a good null experiment, but also provides insight into a probe's susceptibility to

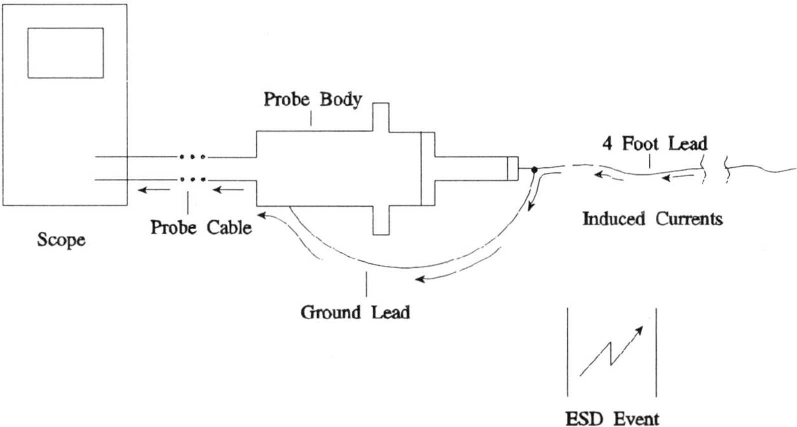

Figure 9.3 Scope probe susceptibility, high impedance 10X passive probe.

interference. At first glance, one might expect a shorted probe to have a zero output, but this is not the case when pulsed EMI is present in the area (and for many other measurement situations as well, see Chapter 4).

In Figure 9.3, current induced by fields of an ESD event several feet away flow in the lead connected to the probe tip, the probe ground lead, and on the *outside* of the probe cable shield. This current can exceed 1 ampere peak. The current causes an $L \cdot di/dt$ voltage drop across the probe ground lead. A value of 10 volts peak is not unusual. In addition to this, the loop formed by the probe and its ground lead can also result in significant pickup if the pulsed EMI event is close by, a few feet away.

In general, the direct connection of *any* lead, including a scope probe ground lead, directly to an EUT *for any purpose* can inject *amperes* of current into the circuit because of the fields of a nearby pulsed EMI event. The injection of high frequency currents such as this into a piece of equipment is very likely to affect equipment operation. Special measurement techniques must be employed to avoid this problem. These will be covered later in this chapter.

Current Probes

Since most current probes do not have to make physical contact with the circuit being measured, they seem like a possible way to make measurements in the vicinity of pulsed EMI. Within limitations to be discussed shortly, they do provide a good measurement option for a pulsed EMI environment.

Current probes, as discussed in Chapter 8, develop an output voltage proportional to the current flowing through them by responding to the magnetic fields generated by the current. The output voltage is related to the current passing through the probe by the transfer impedance, Z_t, which is simply:

$$Z_t = V_{out}/I_{measured} \qquad (9.7)$$

Of the many types available, the hinged donut design is very useful for temporary installation since the conductor need not be cut or disconnected to install the probe. A sample of the types available

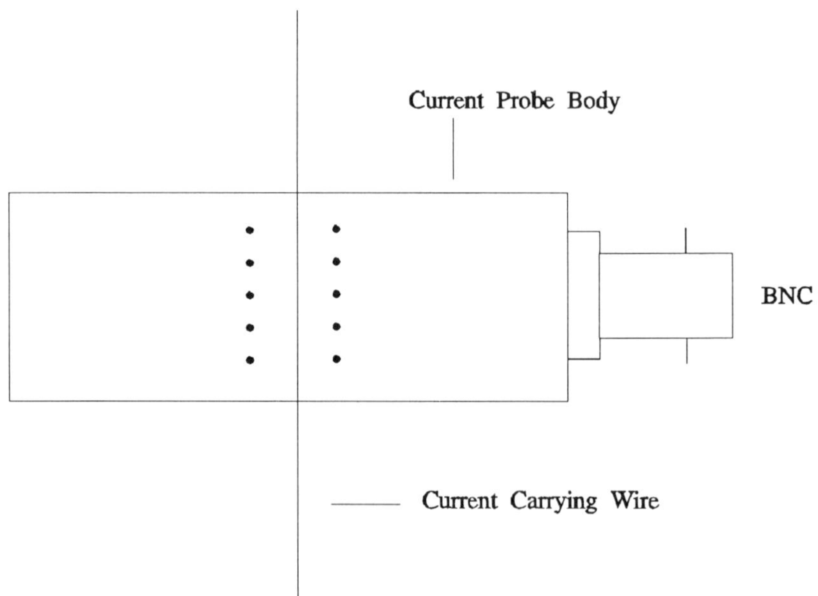

Figure 9.4 Typical current probe use.

were shown in Chapter 8. Figure 9.4 shows a hinged donut probe installed on a wire.

The conductor carrying the current to be measured passes through the hole in the center of the probe. The magnetic field generated by the current in the wire causes magnetic flux to arise in the magnetic core of the current probe which in turn induces a voltage in a pickup coil inside the probe body. (More details on current probe operation are given in Chapter 8.) If the probe is not adequately shielded, electric field coupling from the wire being measured or other sources will cause an output from the probe that is not caused by the current flowing through it but arises instead from the probe's electric field response.

All current probes have an electric field response. Hopefully it is small compared to its magnetic field response. However, since some pulsed EMI events, especially ESD events, can be accompanied by very large electric field changes, it is imperative to check the electric field response of a current probe that is used near pulsed EMI. The electric field, E field, response of a current probe must be checked whether or not the probe is being used to measure current from a pulsed EMI event anytime one shot transient measurements are being made with a storage scope.

Many current probes were designed for use in low impedance, continuous wave (CW) applications. Two examples of these applications are measuring radio frequency current from a transmitter and measuring digital logic noise induced common mode current on a cable with a spectrum analyzer in the frequency domain. Many measurement situations involving pulsed EMI, especially ESD, result in low current in conductors relative to the local electric field caused by the pulsed EMI event. Such a condition can cause some current probes to give an unreliable output.

Current probes with response to DC, such as Hall effect probes, can respond to electrostatic fields and must be used with care. A nearby electrostatic field may erroneously appear as DC flowing through the probe.

Null experiments should be used to verify current probe operation. Two such null experiments are shown in Figure 9.5. The idea of this experiment is to expose the inside of the current probe, or current clamp as it is sometimes called, to the electric field of the wire carrying the current to be measured, but not to allow any *net* current to flow through the probe. Doubling the wire back on itself through the probe accomplishes this for the hinged donut probe in Figure 9.5a. The folded wire should extend into the probe so that the folded end is just flush with the side of the probe opposite to the side from which the wire is inserted. The output of the current probe for the configuration of Figure 9.5a, ideally zero, is the margin of error for the measurement. See Experiment 9.1 at the end of the chapter.

Many current probes are made for permanent installation on a wire. Often, these probes have no hinge and some have a very small center hole that is just large enough for a single wire to pass through. Such an arrangement prohibits the wire from doubling back in the probe, as shown in Figure 9.5a. For this type of probe, usually used on a single conductor, the null experiment can be performed by soldering a small wire stub on the current carrying wire perpendicular to the current carrying wire. The probe is then placed around this stub. The stub should not extend beyond the probe, but just be flush with the probe body. This is illustrated in Figure 9.5b.

Accuracy of Measurements Near Pulsed EMI

Because of the high field strengths associated with pulsed EMI, it is imperative to validate all measurement results with null experiments. The importance of this cannot be overstated. Remember, most measurement apparatus, including instrumentation as well as probes, were

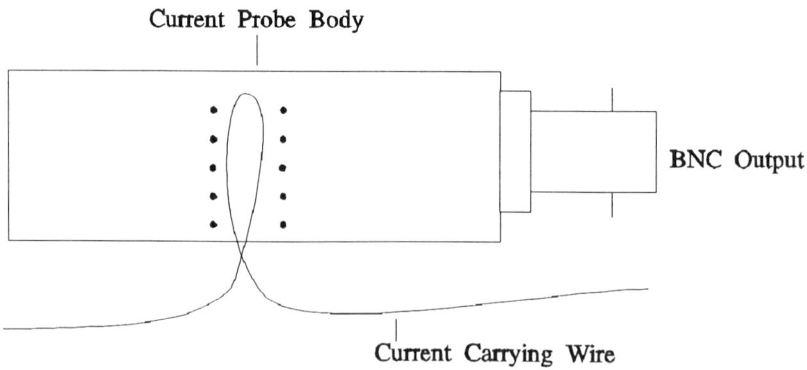

Figure 9.5a Null experiment for a hinged donut current probe.

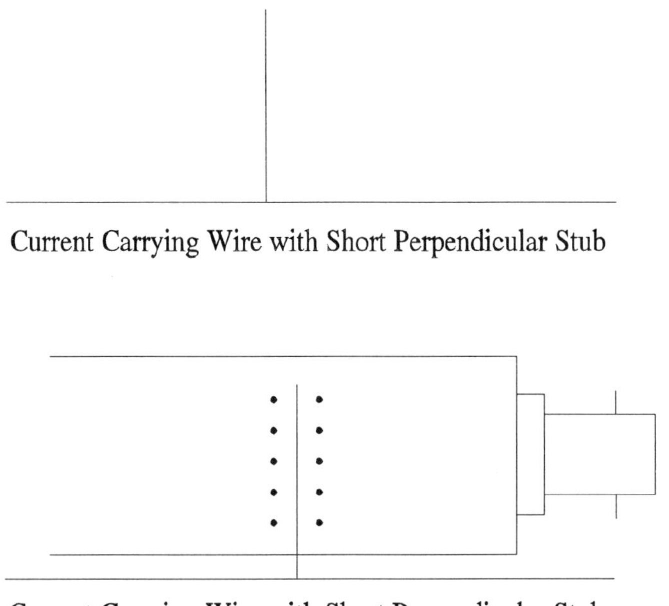

Current Carrying Wire with Short Perpendicular Stub

Current Carrying Wire with Short Perpendicular Stub
Extending into Probe

Figure 9.5b   Alternate null experiment for a single conductor.

not designed for an environment with such strong interfering electric and magnetic fields. It is also imperative to determine what effect, if any, the measurement has on the operation of the equipment under test.

Effect of a Current Probe on the EUT

The effect of using a scope probe or other direct connection measurement equipment to the equipment under test is easily understood. If a current probe is attached to the direct connection, a current on the order of an ampere would not be an unusually high reading to be caused by the fields of an ESD event several feet away. It is not difficult to see that injecting this amount of high frequency current into a piece of equipment can certainly affect its operation. But what effect on a circuit to be measured does a current probe have without a direct connection to the circuit? The answer is that a substantial effect is possible. Luckily, the effect can be measured and minimized.

A current probe can have several pF of capacitance to a measured circuit. Although small, this capacitance can inject significant amounts of high frequency current into the circuit. Remember the example earlier in this chapter where an ampere of current flowed through only one picofarad of capacitance? The current injected into a circuit through the capacitance of the current probe can be measured with a second current probe as shown in Figure 9.6.

The second current probe shown in Figure 9.6 is used to measure the "common mode" current injected into the circuit being measured by the first probe. The distance between the two current probes should be minimized to reduce radiated sources of error and error caused by phase shifts between the two current probes. Under some circumstances, the second probe can indicate that as much as a few

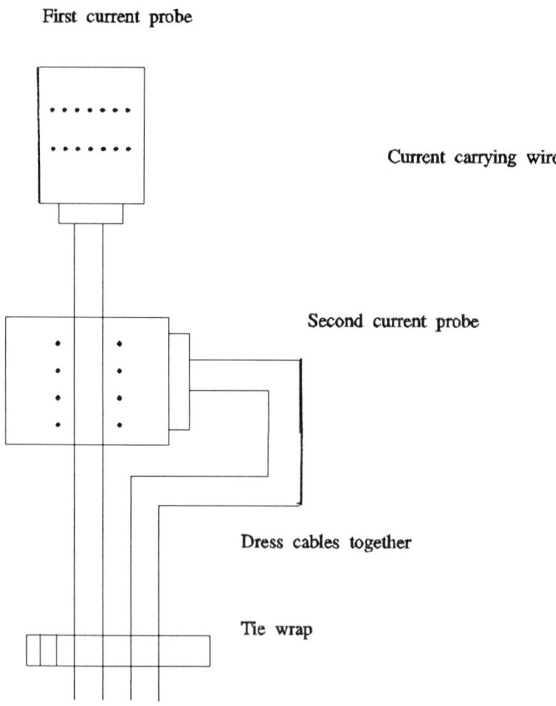

Figure 9.6 Using a second current probe to measure effects on equipment.

hundred milliamperes of current is being injected into the measured circuit by the stray capacitance of the first probe.

The two probe cables should be dressed close together to minimize the loop area between them and thus minimize the voltage between the current probe bodies. Ideally, doing this will reduce the current flowing between the first probe's cable and the second current probe body through the capacitance between them to zero by eliminating the potential difference between the second probe and the cable of the first probe.

The probe bodies should not touch each other or any other conducting surface, if the probe's bodies are conductive. If they do, current may flow in the probe bodies and contribute error to the measurement as well as affect equipment operation. Also, if they touch, increased currents will be set up on the outside of the shields of the coaxial cables that can affect the measurement because of the shield transfer impedance of the coaxial cable.

REALISTIC OPTIONS FOR SYSTEM LEVEL PULSED EMI MEASUREMENTS

There are several methods for making voltage and current measurements in electronic systems in the presence of pulse EMI or other extreme forms of interference. With proper care and design, probes to measure voltage and current can yield accurate results. Probably the best way to make this type of measurement employs optical coupling.

Optical Coupling

Optical coupling represents the best option technically because:

1. There is no conductive path between the EUT and the measurement equipment, and
2. The measurement equipment can be remotely located from the EUT.

These two advantages together minimize both the interference to the measurement and the effect of the measurement on the EUT.

Generally the use of optical coupling involves building small battery powered voltage or current to optical transducers to measure system response and connecting them by optical fiber to optical to electrical convertors at the measuring equipment.[3] Do not assume that just because the electrical to optical convertors are electrically floating and battery powered that they are immune to the effects of interference. Null experiments must always be done to validate measurement data, especially near pulsed EMI. The null experiment for voltage measurements usually involves shorting the inputs of the electrical to optical convertor together at one point on the circuit under test, first to one of the nodes to be measured and then to the other. The null experiment for a current measurement involves the connection of the electrical to optical convertor to the circuit in a way that exposes the convertor to the electric field at the point in the circuit where the measurement is to be made, but with zero current flowing through the probe. Figure 9.5 illustrates the principle as applied to current probes.

Manufacturers of satellites measure ESD effects on satellites using optical methods. Typically, many small battery powered voltage or current to optical transducers are positioned throughout the satellite for the test, and the bundle of fibers carrying measured data is run as much as hundreds of feet to remote measuring equipment.[4]

However, the measurement system and many of its pieces are either not commercially available, expensive, or just not available quickly when the need for this type of measurement arises. The remainder of this section addresses pulsed EMI system level measurements using readily available components that are commercially available, inexpensive, and/or are easily built in a few minutes in the

---

3.   For more information see: "Electrostatic Discharge (ESD) Response Measurements in a Satellite Environment," G. H. Tucker, *1987 EOS/ESD Symposium Proceedings*, EOS-9, pp.124-128.

4.   See Footnote 3.

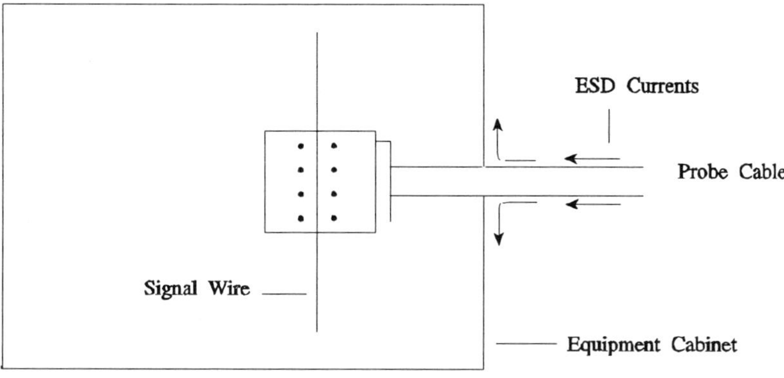

Figure 9.7 Connection of current probe cable shield to equipment cabinet.

laboratory. Measurements made this way have adequate performance for most uses.

Current Probes and Pulsed EMI Measurements

Current probes, properly used, can provide many of the same benefits as optically coupled measurements at lower cost and the probes are easily available in the commercial market. Current probes feature:

1. Reasonable isolation of the measurement equipment from the EUT, and
2. The measurement equipment can be remotely located from the EUT.

But, current probe cables extending into cabinets can bring in interference from outside of the equipment unless precautions are taken. The good news is that the precautions are simple and effective in many cases. In addition, the affect that the current probe is having on the equipment is measurable by a second probe as shown in Figure 9.6.

Dumping pulsed EMI induced currents flowing on the *outside* of the probe cable shield so that they do not penetrate into the equipment is very important. It is easily accomplished, in many cases, by using the skin effect discussed earlier. Figure 9.7 illustrates the principle.

In Figure 9.7, a current probe is used to measure the current in a conductor inside a piece of equipment with a metal enclosure. The current probe cable shield is attached securely in a 360 degree connection to the equipment enclosure. When this is done, ESD or other pulsed EMI induced currents flowing on the cable shield will *not* penetrate the cabinet to a significant extent. This is similar to the flat plate example discussed earlier in this chapter. In that case, current flowed over the surface of the plate taking a path about a hundred times longer that if it had penetrated the 1/8 inch plate directly.

A 360 degree shield contact to the equipment enclosure is desirable, but any contact helps. Use a second current probe to see how much of the pulsed EMI current penetrates the equipment enclosure. Doing this is especially important if a 360 degree shield contact is

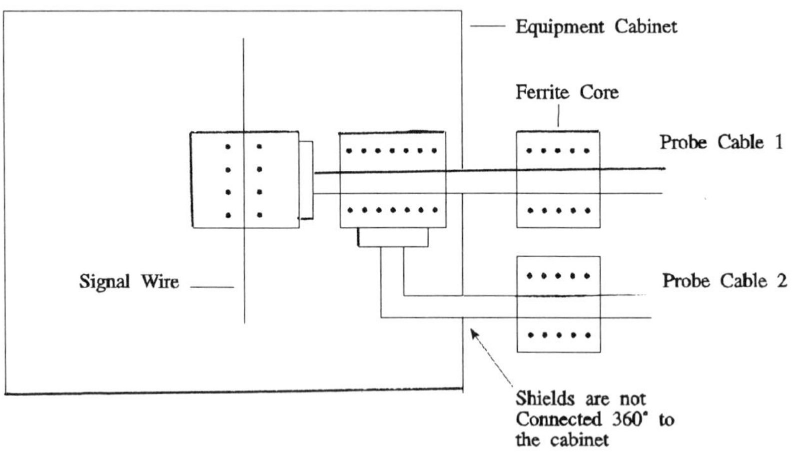

Figure 9.8 Partial shield connection example.

not possible, or if the cabinet has holes or seams in the vicinity of the probe cable entrance.

Figure 9.8 shows an example of measuring the current on a cable shield that penetrates a cabinet enclosure using a second current probe. Remember that the probe bodies, if conductive, should not touch each other or other conducting surfaces inside of the equipment. In addition to possible errors induced into the measurement, if either of the probes touches a conductive surface inside the equipment there will be a direct conductive path into the equipment increasing the shield current that is injected into the circuit. This is especially true if the probe cable shield cannot be securely fastened to the enclosure. The second probe, measuring the shield current of the main probe, should be located as close as possible to the enclosure surface to measure how much pulsed EMI current is penetrating the enclosure. Alternately, if the second probe is located close to the main probe, an upper bound for the current flowing through the capacitance between the main probe and the circuit under test is measured.

The amount of shield current flowing on the main probe cable in Figure 9.8 can be reduced by the addition of the ferrite cores shown in the figure. These cores should be made of lossy material of the type used in EMI mitigation. If the cores were made from low loss material, such as might be used in a high frequency transformer core, only inductance would be added to the circuit by the cores. The added inductance might resonate with other circuit reactances and cause an undesirable response. Common mode shield currents on the cables being treated could even be increased in some instances.

A lossy ferrite core placed around a cable effectively adds an inductance in parallel with a resistance as a series impedance in the cable to reduce common mode currents. A typical core may inject a resistance of 100 ohms in parallel with an inductance having a corner frequency of about 50 MHz. Although several turns of a wire or cable can be put on a ferrite core to increase the series impedance, one must be careful of the resulting interturn capacitance. This capacitance can compromise the core's performance at high frequencies by bypassing the impedance that the core injects into a circuit.

Using ferrite cores to reduce common mode current on current probe cables is a handy technique when the cable shields cannot be grounded well, or not at all for that matter, to an enclosure. In any event, the second probe will measure how much current is being injected into the inside of the equipment.

The amount of current getting into the equipment that is allowable is dependent on the type of equipment being measured. If the equipment contains only digital logic, currents that result in an $L \cdot di/dt$ drop of less than a few tens of millivolts per inch on a wire are not likely to cause problems. If sensitive circuitry is present in the equipment, such as an optical receiver that works on very small signal levels, the $L \cdot di/dt$ drop of the interfering shield current may have to be reduced much further.

Using Pairs of Current Probes

The usefulness of pairs of current probes is illustrated by the measurement of parasitic common mode shield current penetrating the cabinet in Figure 9.8. However, pairs of current probes can be used in many other ways to yield useful information. Examples to be discussed in this section include making null measurements simultaneously with current measurements, and measuring/using the phase between two currents.

Using current probes in pairs yields other important benefits. Increased measurement speed and accuracy can result. For instance, measuring parasitic common mode current helps to predict the effect of the measurement on the system. Being able to do a null experiment simultaneously with a current measurement increases confidence in each measurement. If only one current probe is used, either the measurement speed is reduced by the need to do null experiments or the measurement accuracy is suspect.

Phase information is a very powerful tool. By knowing the relative phase between two currents, one can better understand the coupling mechanisms of pulsed EMI and more easily design a fix as discussed later in this chapter.

To illustrate simultaneous null experiments consider the example shown in Figure 9.9. In the figure, each current probe is connected to a vertical input of a dual trace digital storage scope. Since the two current probes are physically reversed on the wire, the two traces on the scope will be mirror images of each other, one trace being the

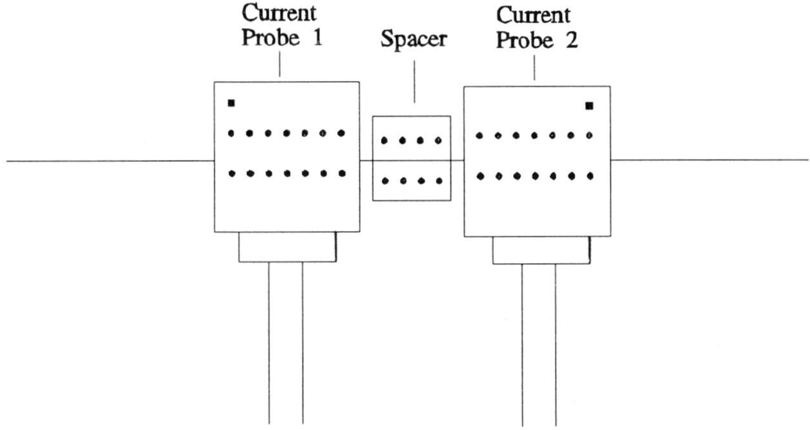

Figure 9.9 Simultaneous null experiment using two current probes.

inverse of the other, *provided* there is no common mode interference to the measurement and the probes are very close to each other.

Common mode interference will result in the two traces not appearing as mirror images. For instance, capacitive coupling from the wire into the probe will make the two traces move in the same direction, both up or both down, whereas a real current will make one trace move up and the other down. Common mode interference can occur in the current probes, the cables connecting them to the scope, or in the scope itself.

Since the bodies of most current probes are conductive, they should be separated by an insulator, such as a roll of tape, as shown in Figure 9.9, to prevent currents from flowing between the current probe bodies. Such currents would flow over the surface of the probes and might contribute to the output of one or both probes if it passes through the middle of either probe, adding error to the measurement. Similarly, as discussed earlier neither probe should be allowed to touch any other conductive surface.

Figure 9.10 shows the resulting scope patterns for two probes configured as shown in Figure 9.9. The source of the current is an

EFT, **E**lectrical **F**ast **T**ransient, burst generator used to test for compliance to IEC 801-4. The current probes used were Fischer F33-1 probes with a 5 ohm transfer impedance so the vertical scale of 2.5 volts per division yields a vertical sensitivity to current of one half ampere per division.

The current waveforms peak at slightly over one ampere of current, actually just saturating the scope near the start of the trace. Notice that the two waveforms are nearly identical mirror images of each other. This is a sign that the measurement is free from significant common mode interference.

Since some current probes were not designed for use near high electric fields, a result like that in Figure 9.10 is not always assured. There are some commercially available probes in which the two waveforms produced by a setup similar to Figure 9.10 would bear little relationship to each other, and are certainly not mirror images. When this happens, the measurement is not to be trusted.

Figure 9.11 shows two probes measuring the relative phase between two currents, possibly caused by pulsed EMI. By comparing the two current probe outputs displayed on a scope, one can determine if the currents in the wires are common mode or differential mode, or a combination of both. Common mode current will look the

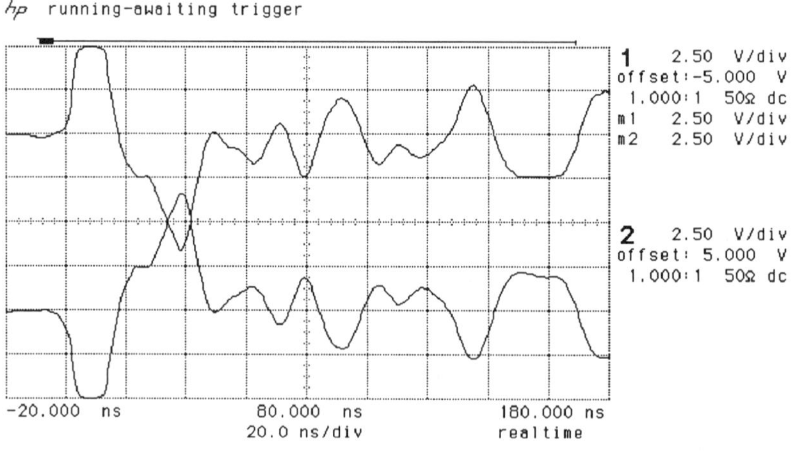

Figure 9.10 Simultaneous current probe null experiment result.

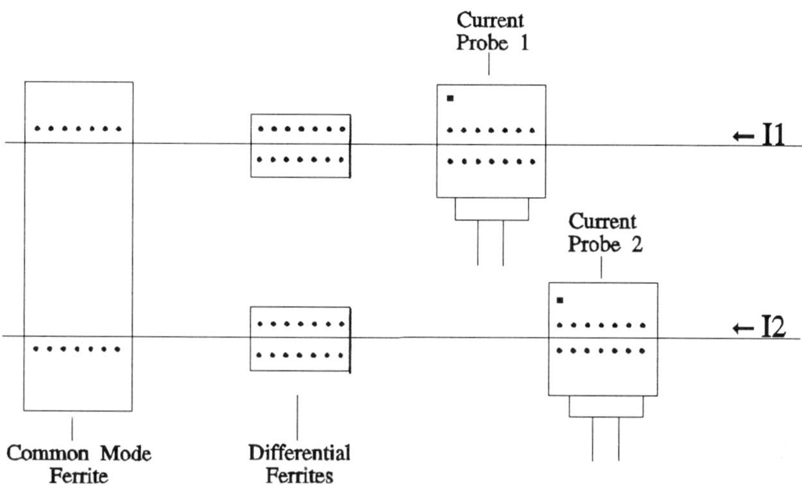

Figure 9.11 Using phase information.

same on both traces, whereas differential mode current will appear as mirror image waveforms similar to Figure 9.10.

If the currents are common mode, one ferrite around both wires will reduce the noise currents without affecting a differential signal. If the noise currents are differential, an individual ferrite around each lead is necessary to reduce the pulsed EMI current.

Figure 9.12 shows a recommended cable for connecting the dual current probes to a measurement instrument such as a digital scope. The cable is similar to that used for the balanced coaxial probe discussed in Chapter 6 in that the shields of the two cables are soldered together periodically. As with the balanced coaxial probe, soldering the shields together periodically minimizes loop area for the induction of parasitic differential noise currents on the shields and helps ensure that the two shield currents are equal and common mode. In addition, ferrite cores are installed on the cable to reduce the parasitic common mode currents induced on the cable by strong fields. The shield of the cable is exposed in a few places near the current probe end to provide a place to dump noise currents flowing on the outside of the cable shields to chassis ground.

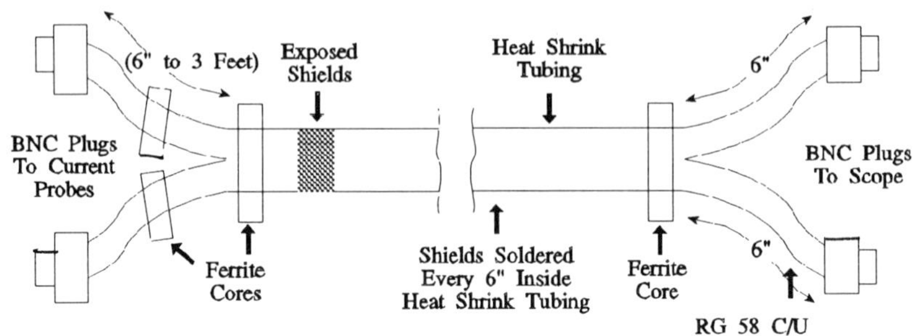

Figure 9.12 Dual current probe cable.

The total cable length may vary from a few feet to several tens of feet depending on equipment size and the need to move the measurement equipment away from the equipment under test. Experience has shown that soldering the shields at about 6 inch intervals gives good performance. High quality, double shielded coaxial cable should be used for best results.

Figure 9.13 shows the experimental setup for a measurement of the shield current on a length of RG58C/U cable that penetrates a metal cabinet about the size of a personal computer. One current probe is located on the outside of the box and the other on the inside. Both probes are located near where the cable passes through the box. The cable shield for the probe on the inside is securely connected the chassis in a 360 degree connection. The RG58C/U cable is terminated in 50 ohms inside the box and its shield grounded to a circuit board at the termination. The circuit board is connected to the metal box at one corner. The source of the noise current on the shield of the RG58C/U for the following figures was an IEC 801-4 EFT generator.

Figures 9.14 through 9.17 show the amount of shield current that penetrated into the box of Figure 9.13 for several conditions. The top traces in the figures are for the external current probe and the bottom traces are for the output from the internal current probe. In Figure 9.14, the cover was off the box, resulting in a U-shaped chassis, and the shield of the coaxial cable was not connected to the chassis as it passed through.

As one might expect, the current inside the box is identical to the current outside the box on the coaxial shield, having almost a 1 ampere peak value and a period of oscillation of about 200 nanoseconds. The period of oscillation of the current represents the natural frequency of resonance of the cable as modified by the close surroundings. The cable was about 40 feet long with part of its length coiled into several turns about 2 feet in diameter.

Figure 9.15 shows the same waveform with a shorter time scale, 20 nanoseconds per division instead of 100 nanoseconds per division. On this time scale, the higher frequency components of the noise current are easily seen. This time scale is used in Figures 9.16 and 9.17 as well.

In Figure 9.16, the box is completely closed and the shield of the RG58C/U coaxial cable is fastened to the box in a 360 degree fashion as the cable enters the box. Notice that the bottom trace of the internal current probe has a vertical sensitivity 10 times that of the top trace to give 50 mA per division.

Even with the increased sensitivity, the current that penetrates the metal box is difficult to see, being on the order of 40 dB lower than the external current. It is obvious that the box makes an excellent

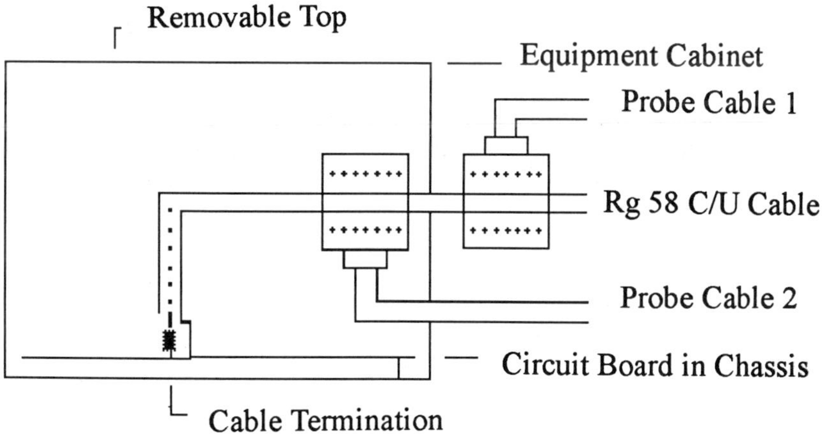

Figure 9.13 Shield current penetration measurement.

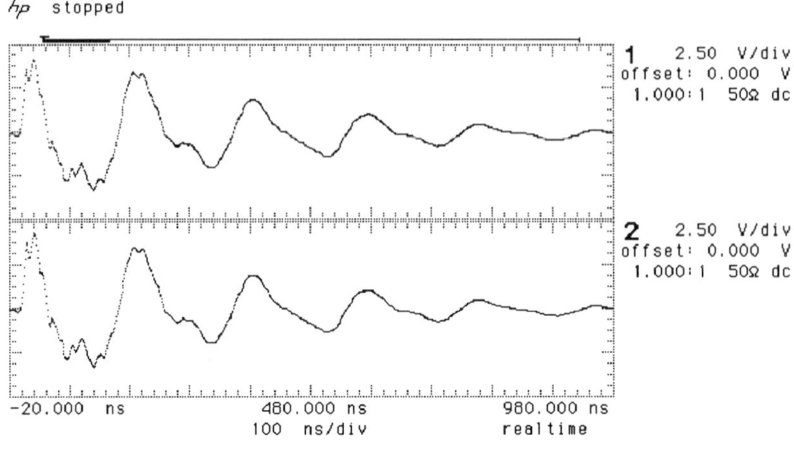

Figure 9.14 Shield current penetration, shield not connected to box (100ns/div).

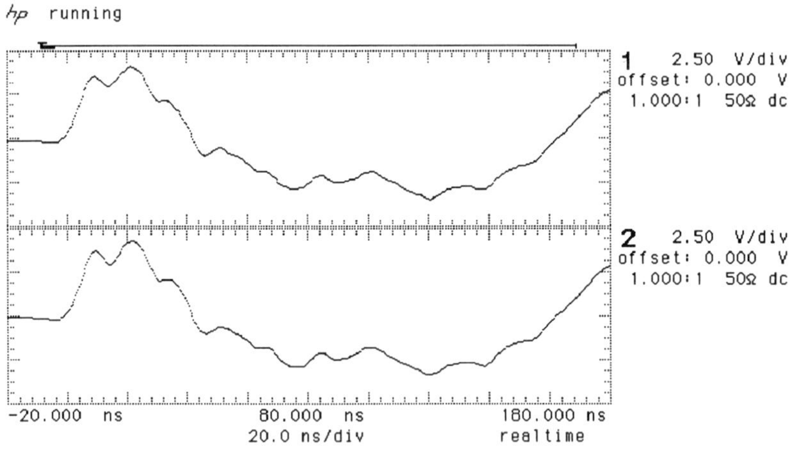

Figure 9.15 Shield current penetration, shield not connected to box (20ns/div).

shield for common mode currents. This is because the skin effect keeps the noise currents on the *outside* of the box.

The waveforms of Figure 9.17 were taken under the same conditions as were present in Figure 9.16 except that the cover of the box is removed leaving a U-shaped chassis with the cable going though it near the middle of one side. Again, the bottom trace has an increased vertical sensitivity.

The removal of the top of the chassis allows some of the cable current to spill over the edge of the box into the interior. In addition, local fields may be inducing current in the loop formed by the cable shield, the circuit board connection to the chassis, and the chassis itself. In any event, the shielding effectiveness of the equipment enclosure, its ability to keep common mode shield currents out, has been reduced to about 30 dB.

This example serves to show how useful measurements can be easily made with pairs of current probes. The exact effect of various shield and box configurations on shielding effectiveness were easily and quickly measured. And, no exotic equipment was necessary to do it, just a pair of high quality commercially available current probes connected to a 1 GSa oscilloscope via the dual cable shown in Figure 9.12.

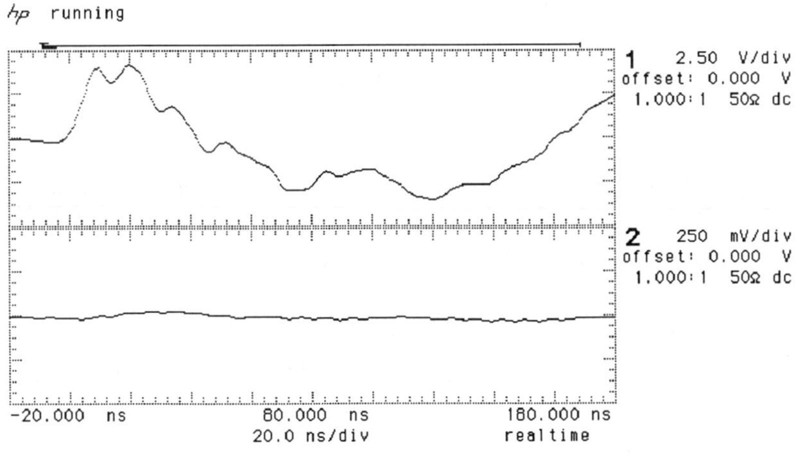

Figure 9.16 Shield current penetration, shield connected to box 360° (20ns/div).

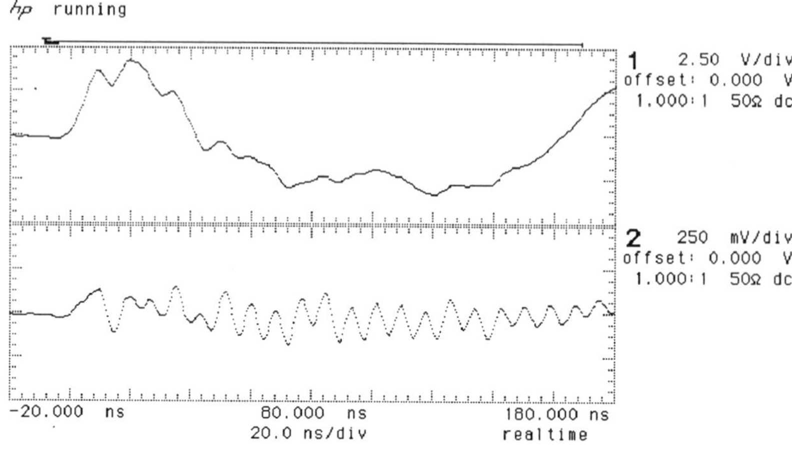

Figure 9.17 Shield current penetration, shield connected to box 360° (20ns/div, box open).

Voltage Measurements and Pulsed EMI

As discussed earlier in this chapter, conventional scope probes are not recommended for measurements around pulsed EMI. However, it is not always necessary to resort to optical coupling either for voltage measurements. Many voltage measurements can be accurately made using a special adaptation of the balanced coaxial probe described in Chapter 6. This probe features:

- Good common mode rejection over a wide frequency range,
- Flat frequency response over a wide frequency range,
- Adequately high input impedance (typically 500 ohms), and
- Easy and rugged construction.

Figure 9.18 shows a diagram of the balanced coaxial probe adapted for pulsed EMI use. The features that differ from the probe discussed in Chapter 6 are:

1. Terminations on the probe ends of the cable. Four 200 ohm resistors are soldered between the center conductor of each coax cable to its shield to form a 50 ohm termination to prevent reflections at the probe tip end of the cables. This causes the impedance seen looking into the cable from the probe tip to be 25 ohms (the 50 ohm cable in parallel with the 50 ohm termination). Thus 475 ohm resistors are used at the probe tip to make a probe with a 500 ohm input impedance. Termination of the cables allow use of:

2. A 180 degree combiner to do the A-B subtraction outside the scope. This eliminates errors caused by the digital subtraction in the scope, and has the added advantage of requiring only one scope input. The combiner is essentially a multiwinding transformer, but it should be terminated in 50 ohms at all ports and thus the terminations in point 1 above. Don't forget to include the combiner loss when calculating the overall probe attenuation factor.

3. The shield of the cable is exposed in a way to permit its connection to the chassis or enclosure of the equipment under test.

When using the balanced coaxial probe for measurements inside of an electronic system in the presence of pulsed EMI, the shield of

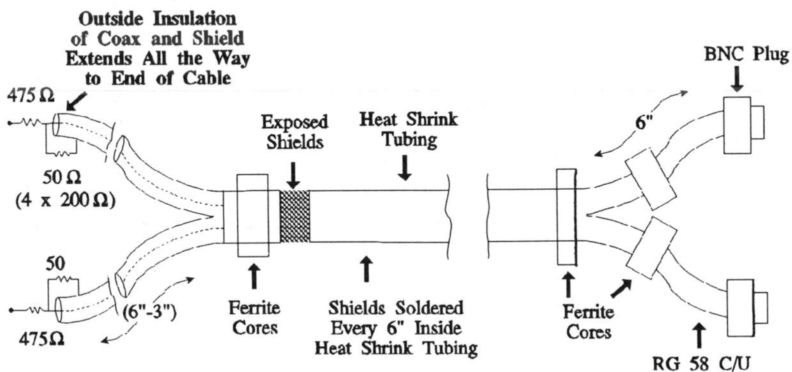

Figure 9.18 Balanced coaxial probe adapted for pulsed EMI.

the probe should be connected to the equipment enclosure or chassis for two reasons. First, to dump pulsed EMI induced shield currents to the chassis and keep them out of the equipment. And, second, to keep the common mode voltage low enough to be able to make the desired differential measurement at the sensitivity required. If further reduction of common mode voltage is necessary, an unlikely event, a ground lead can be added from the end of the cable shield to the circuit ground of the system inside the cabinet at the expense of increasing the amount of pulsed EMI generated current that is injected into the system circuitry. If a ground lead is used, the current injected into the system circuit ground should be checked with a current probe. If the $L \cdot di/dt$ drop, in volts per inch, is too great, reduce it with ferrite cores.

The null experiment for this probe simply requires the connection of the two probe tips (usually the ends of the 475 ohm resistors) together, and then the connection of them to each of the two nodes to be measured, one at a time, in the presence of the pulsed EMI. Any signal measured represents the common mode response of the measurement and is the margin of error for the voltage measurement. This process is illustrated in Figure 9.19.

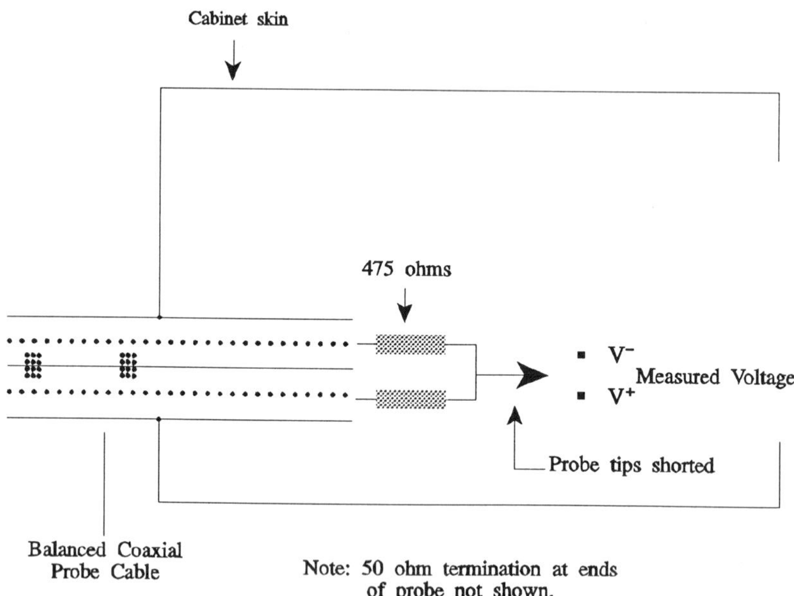

Figure 9.19 The balance coaxial probe null experiment.

The 475 ohm resistors provide additional isolation between the system circuitry and the pulsed EMI induced shield current. If the shield of the probe is not connected to the equipment under test except at the cabinet skin, as shown in Figure 9.19, then any pulsed EMI currents that penetrate the cabinet, because of less than optimal shield connection to the cabinet, will still have to pass through the 475 ohm resistors to get into the internal circuitry. Use a current probe to measure just how much EMI current is getting into the equipment.

Figures 9.20 through 9.23 show examples of measurements using a balanced coaxial probe on the box previously described in Figure 9.13 for the current probe measurements of shield currents. In this case, the box was closed and the probe was measuring the differential voltage across a balanced termination inside the box. A balanced cable entered the box and was terminated by the balanced termination. An IEC 801-4 EFT burst generator was used to induce noise on the balanced cable.

Figure 9.20 shows the voltage measured for one configuration of balanced cable, and Figure 9.21 shows the null experiment result with the probe tips shorted together and connected to one side of the balanced load. The probe attenuation factor combined with the combiner loss results in a vertical scale of 200 mV/div, not the 12 mV/div displayed on the scope readout in these figures.

In Figure 9.20, the waveform reaches a peak value of 800 mV with risetimes of about 4 to 5 nanoseconds. The questions to be asked are whether the observed waveform is accurate, and how much effect the measurement is having on the system circuitry? These questions are answered by the null experiment shown in Figure 9.21.

In Figure 9.21, the common mode response is only about 150 mV, the error in the measurement, and has risetimes that are much slower than the measurement of Figure 9.20. This much common mode response is low enough to not significantly affect the accuracy of this measurement.

While it is possible that dumping the shield currents of the probe on the outside of the equipment enclosure could effect the response of the system to pulsed EMI, the effect is usually weak and is not likely to noticeably affect the equipment operation. This is especially true since most equipment would have several other cables attached to it, all contributing noise currents.

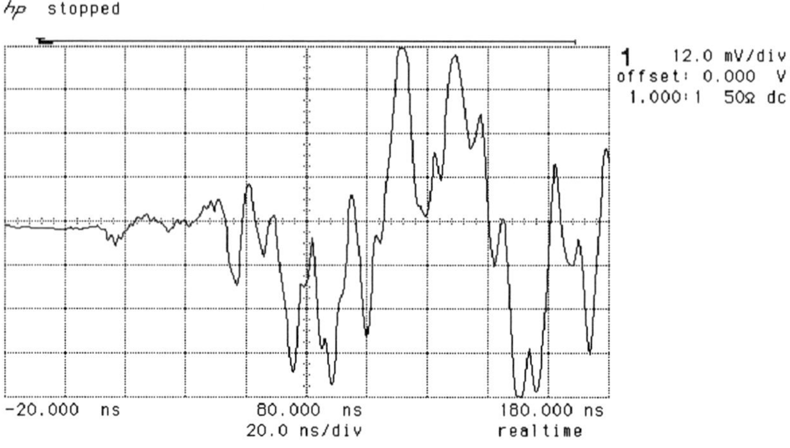

Figure 9.20 Measured differential voltage, first configuration (200mV/div).

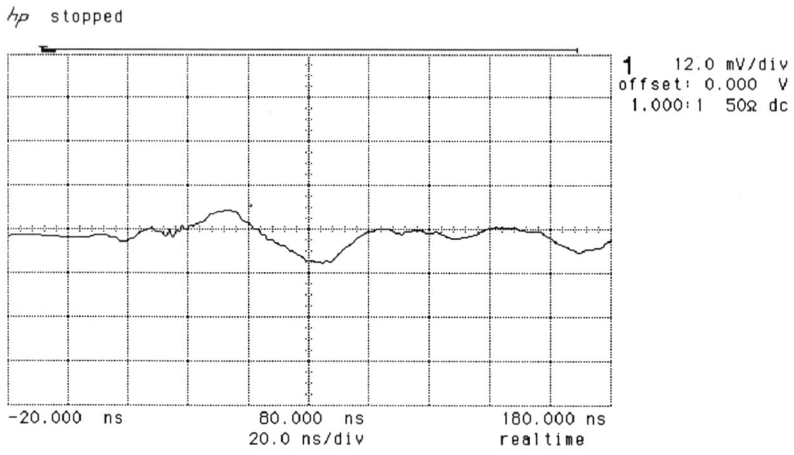

Figure 9.21 Null experiment, first configuration (200 mV/div).

Figures 9.22 and 9.23 show the differential signal and null experiment results for another cable and equipment configuration similar to the last one.

For this configuration, the differential signal is just under 300 mV peak and the null experiment result is about 100 mV. The common mode response is a much greater fraction of the measured differential voltage and could lead to an error of almost 50 percent in the measurement in this case.

The Coaxial Current Probe

A coaxial current probe is built using a length of coaxial cable with a special termination at the probe end. The principle has been used to measure high frequency currents for many years by ANSI, and by the IEC to calibrate ESD simulators, although in a more complex form to achieve higher bandwidth. This current probe has a transfer impedance of 1 ohm and is useful for calibrating commercial current probes. It is easy to build, and bandwidths of several hundred MHz

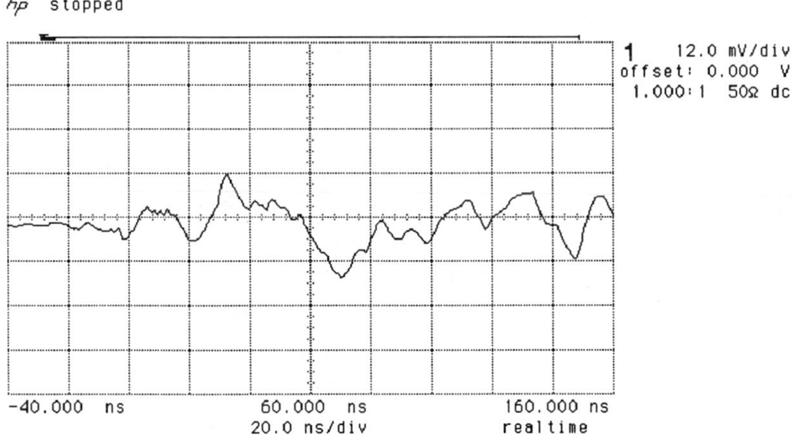

Figure 9.22 Measured differential voltage, second configuration (200 mV/div).

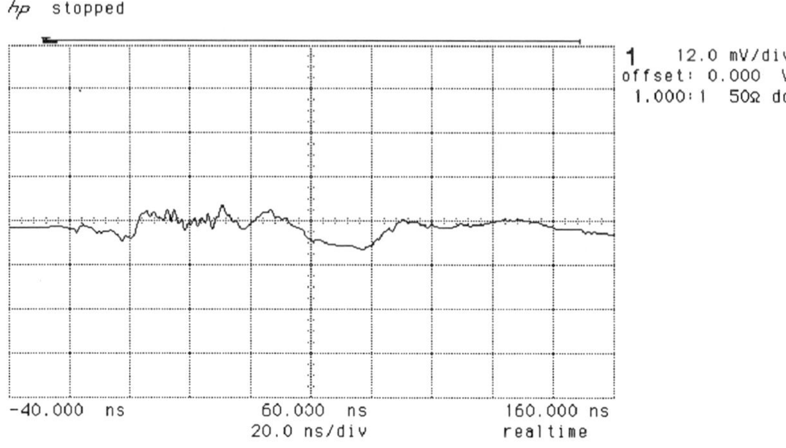

Figure 9.23 Null experiment, second configuration (200 mV/div).

are relatively easy to achieve. Figure 9.24 shows the principle of operation.

Figure 9.24a shows a simplified circuit of the probe. A coaxial cable is terminated in its characteristic impedance of 50 ohms at an oscilloscope or other instrument. At the other end of the cable a 47 ohm resistor is connected to a wire and a 2 ohm resistor is connected from that wire back to the cable shield.

The equivalent circuit of Figure 9.24b can be derived in the following way. Looking into the coaxial cable from the 47 ohm resistor, one sees 50 ohms so the whole circuit inside the coax cable, including the scope termination, can be replaced by a 50 ohm resistor. The outside of the shield of the coax cable must still be modeled, and is shown as a wire back to the scope chassis in Figure 9.24b.

A current, I_1, flowing on the wire will see a total impedance of 97 ohms in parallel with 2 ohms on its way to the shield of the cable and ultimately the scope chassis. Some of the current will radiate from the shield so the amount of current reaching the scope chassis will not necessarily be the same as the current which started out on the cable shield, particularly for higher frequencies above a few tens of MHz. The parallel combination of 97 ohms and 2 ohms is approximately 2 ohms, so 1 ampere of current will produce about 2 volts. About one half of this 2 volts will appear across the 50 ohm load and be sent down the inside of the coax cable to the scope. Thus the scope will display about 1 volt per ampere flowing through the 2 ohm resistor giving a transfer impedance of 1 ohm.

The inductive reactance of a 2 ohm resistor may exceed its resistance at few tens of megahertz. To extend the accurate bandwidth of the probe into the hundreds of megahertz, the 2 ohm resistor should be composed of several higher valued resistors in parallel, equally spaced around the coaxial cable as shown in Figure 9.25. In this case, five 10 ohm resistors make up the 2 ohm resistor. All of the resistors should be $\frac{1}{2}$ watt carbon composition resistors. Carbon composition resistors are less inductive and tolerate high pulsed currents better than do most film type resistors.

The coaxial current probe has several uses in the measurement of pulsed EMI effects on electronic circuits. One is for checking a commercial current probe's transfer impedance and suitability for measuring the sometimes high peak currents that pulsed EMI can generate. This can be done by connecting the tip of the coaxial current probe to a piece of metal through a few feet of wire. Position the commercial current probe around the resistor network of the coaxial current probe and connect both probes to the vertical inputs of a digital storage scope with a sampling rate of at least 1 gigasample

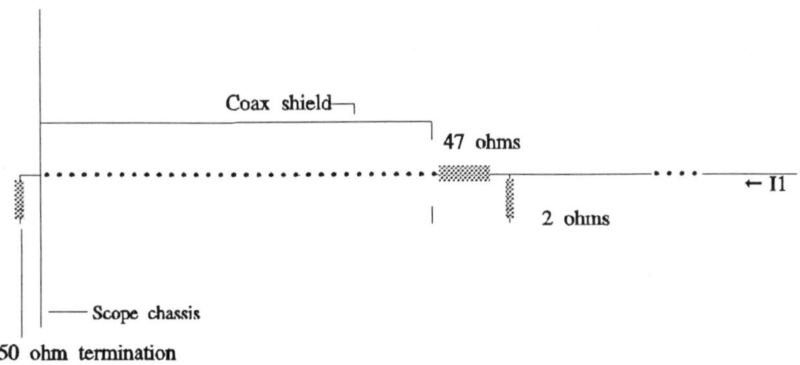

Figure 9.24a Coaxial current probe (simplified).

per second. Comparison, on the scope screen, of the outputs of the probes induced by a nearby ESD event should show identically shaped waveforms, if the commercial probe has adequate bandwidth and is not saturating from too much current. The relative amplitudes of the two traces on the scope gives the transfer impedance of the commercial probe since the coaxial probe has a 1 ohm transfer impedance.

This coaxial current probe can also be used to check shielding effectiveness of an enclosure against pulsed EMI induced shield currents on a cable without requiring the use of standard current probes. This technique is shown in Figure 9.26.

The coaxial current probe tip is connected to circuit ground inside a piece of equipment. Its shield is connected to the equipment metal enclosure by different methods to be evaluated, such as a pigtail or directly in a 360 degree fashion. A few thousand picofarad capacitor in series with the probe tip can be used if DC isolation between the circuit ground and equipment enclosure is required.

With the probe's output displayed on a digital scope, the amount of pulsed EMI current entering the equipment enclosure is easily seen, and the effect of different enclosure and cable shield grounding arrangements can be judged. This arrangement simulates an adjunct device, the scope in this case, connected by a shielded cable to the equipment under test.

ESD event detection and logging can also be accomplished using this probe and a digital scope. The detection and logging of ESD and

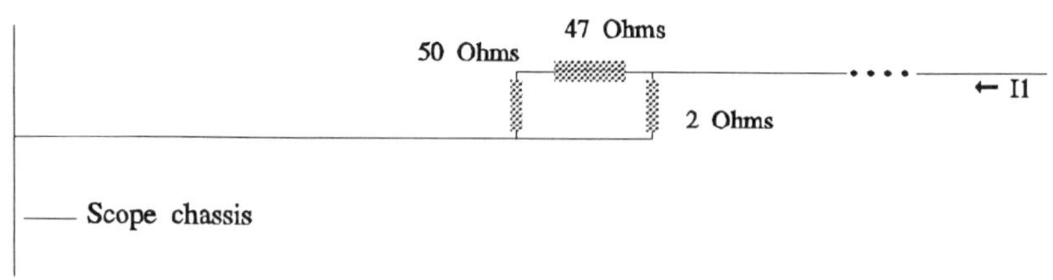

Figure 9.24b Coaxial current probe equivalent circuit.

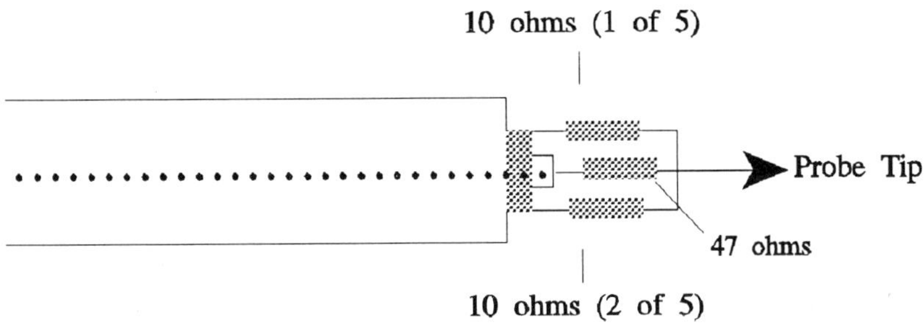

Figure 9.25 Coaxial current probe construction.

other pulsed EMI events is useful in troubleshooting equipment field problems, to see if they are caused by pulsed EMI. It is also useful in evaluating an environment for the severity of pulsed EMI and its effect on equipment not yet installed. Another possible use is for monitoring an ESD controlled area for unwanted discharges.

Figure 9.27 shows the use of the coaxial current probe for logging pulsed EMI events. Pulsed EMI event logging can be accomplished by connecting the probe output to the 50 ohm input of a digital scope and by connecting a 3 to 6 foot wire to the probe tip. Nearby pulsed EMI will generate electromagnetic fields that cause current to flow in the wire and probe cable shield. By adjusting the scope trigger level, one can control the intensity of the pulsed EMI event that will trigger the scope. It is also useful to connect a counter to the trigger gate output of the scope for counting events. Recorded values of current range from a few amperes for nearby—a couple of feet away—severe ESD events to tens of milliamperes for small ESD events several tens of feet away. This test arrangement will also be sensitive to EFT, Electrical Fast Transient, events described in IEC 801-4. The amplitudes will be somewhat less than for strong ESD.

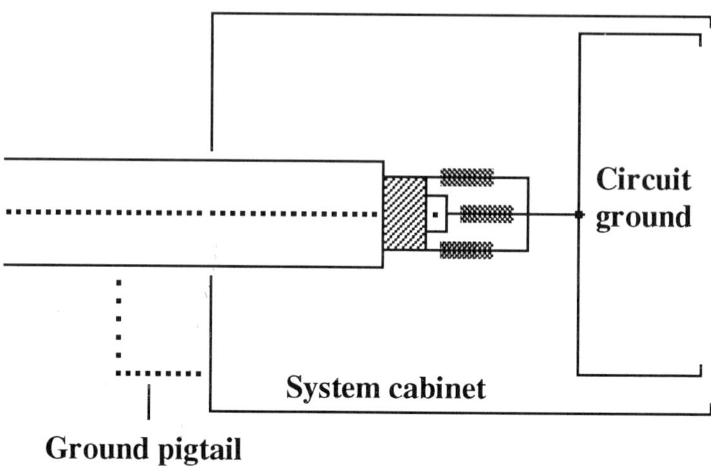

Figure 9.26 Shielding effectiveness measurement.

SYSTEM LEVEL ESD AND EFT MEASUREMENTS VERSUS TRIAL AND ERROR

The typical sequence of events for a system level ESD test, or other pulsed EMI test as well, is:

1.   Design and build system,
2.   Test system for ESD response,
3.   Fail ESD test,
4.   Add on lots of ESD fixes to system,
5.   Retest and fail again,
6.   Add more fixes, and
7.   Finally, pass the ESD test.

The above sequence can be time and resource consuming. And when the fix is achieved, what was the margin of the fix? Were all the fixes installed necessary? Measurement of the effect of pulsed EMI on an electronic system often allows the designer to pinpoint the problem and apply an efficient fix. This process can be done in

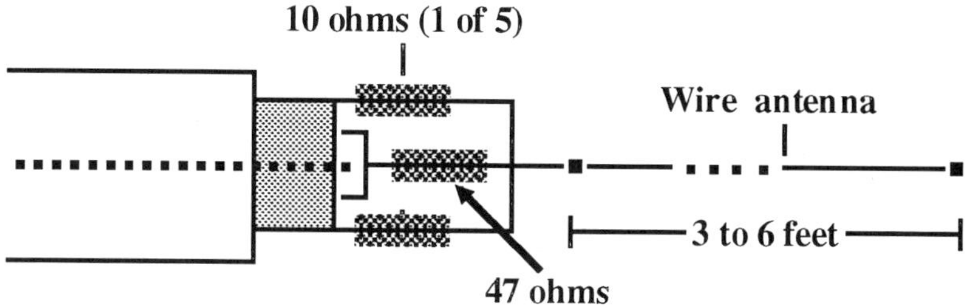

Figure 9.27 EDS event logging example.

a development laboratory before the equipment is sent out for official compliance testing.

The measurement techniques covered in this chapter also allow for subthreshold testing. Pulsed EMI levels much smaller than those required for system malfunction can generate easily measured signals. This allows a system to be analyzed for a problem and a fix to be designed without the need for system upset or damage. Subthreshold testing is useful for installed equipment that must remain working at all times, such as that in telephone central offices.

The most challenging type of equipment to measure is equipment with a nonconducting enclosure or no enclosure at all. Finding a good place to dump pulsed EMI induced shield currents can be difficult. The use of ferrites and current probes becomes more important for this type of equipment as opposed to the balanced coaxial probe with its direct connections.

SUMMARY

Pulsed electromagnetic interference (including ESD and EFT) has become important to designers of electronic systems in recent years. This chapter has discussed how to measure voltages and currents inside an electronic system exposed to pulsed EMI. Pulsed EMI can generate large voltages and currents because of circuit parasitic capacitance and inductance. The interference potential is so great that one always must be suspect of measurements made near pulsed EMI.

There are two main tools easily available to engineers for making these types of measurements. They are the current probe and the balanced coaxial probe. These probes can give accurate results if these points are followed:

- Dump shield noise currents to chassis or enclosure ground,
- Measure the effect of the measurement on system operation (a second current probe is useful for this), and
- Be skeptical and do null experiments.

The coaxial current probe can be a useful addition to the "tool box." This probe can be used to calibrate a commercial current probe and as a detector for pulsed EMI events in an area. It can be built in a few minutes using a few resistors and a length of coaxial cable. Using it to characterize the pulsed EMI environment (for frequency and intensity of events) can provide valuable information to systems designers and field support personnel.

Using the techniques presented in this chapter, one can break the cycle of the trial and error approach to system level pulsed EMI troubleshooting. The advantages of the measurement approach over the trial and error approach are:

- Reduced testing and fix time,
- Increased confidence in the fix (knowing the margin of the fix),and
- Subthreshold testing.

These techniques have successfully found and fixed many system pulsed EMI problems over the years. In some instances, problems that had defied the trial and error approach for weeks or months were fixed quickly, in less than a day.

LABORATORY DEMONSTRATIONS

EXPERIMENTS

Equipment needed:
- Two current probes with a bandwidth from a few MHz to about 200 MHz. (Fischer Custom Communications F33-1 probes or equivalent),
- Contact mode ESD simulator as per IEC-801-2, revised,
- Dual channel digital oscilloscope with a sampling rate of 1 GSa/sec or more,
- Dual current probe cable shown in Figure 9.12,
- 10 feet of insulated wire with the insulation removed for about ¼ inch in the middle, and
- A 1 foot square piece of metal.

Experiment 9.1: Current Probe Measurements

1. Connect one end of the 10 foot wire to a large grounded metal object such as a file cabinet, the green wire ground pin of an electrical socket, building frame, metal wall, etc. Lay the wire out straight on a wooden table with the exposed wire at the midpoint of the wire on the table. Fasten the other end of the wire to the ESD simulator tip.

2. Connect the current probes to the 50 ohm inputs of a digital oscilloscope using the dual current probe cable. Display each probe's output on a separate trace of the scope such that the traces will not overlap when the waveforms are displayed. Set the scope sensitivity to display a full scale voltage of 20 volts peak for each trace without overlap of the traces or saturating at the top or bottom of the screen. Set the scope trigger for a positive edge of 2 volts on channel 1.

3. To check the probes for spurious response put the wire through one probe (channel 1, on the side closest to the ESD simulator) and looped into the other as shown in Figure 9.5. The probes should be close to each other and near the center of the wire, but not touching each other or anything conductive. Generate a

contact discharge of about 2 kV from the ESD simulator and note the waveforms on the scope.

4. Now put the second current probe (channel 2 of the scope) around the wire such as to be about 1 inch from the first probe and reversed with respect to the first probe on the wire. The exposed wire at the center of the 10 foot length should be between the two probes. Generate another 2 kV contact discharge and note the waveforms on the scope.

5. Reverse the second probe on the wire so that the two probes are in phase on the wire and separated by 1 inch with the exposed conductor between them. Generate another contact discharge and note the waveforms.

6. Repeat step 5 with the 1 foot square metal plate supported by books or other nonconductive supports such that the plate is perpendicular to the wire and just touching the exposed wire between the current probes. Generate another 2 KV contact discharge and note the waveforms on the scope screen.

Questions:

9.1.1   How should the two waveforms compare in step 3?

Answer: The probe with the wire passing through it will register a current on the order of 4 amperes peak with a ristime of less than one nanosecond (one sample by the scope). The rise will be followed by a relatively low frequency ringing of a few tens of MHz. that represents the natural resonant frequencies of the 10 foot wire and its ground connection. The second current probe is being used to perform a null experiment. Since no current actually passes through the probe, its output should be zero. However, most current probes will have some spurious output. In order to be useful for these kinds of measurements, the spurious output must be well below the output due to the current to be measured.

9.1.2   How does the output of the second probe change when the looped wire is pulled through the probe so that it extends beyond the probe up to a foot?

Answer: While the current still passes through the probe and back again, the net current through the probe may increase from zero as the loop is pulled further through the probe and cause an increasing output. This is due to phase shift in current along the wire because at the frequencies involved, a foot or two of wire represents a significant fraction of a wavelength.

9.1.3   How do the waveforms from the two probes compare in steps 4 and 5?

Answer: Since the two probes are placed around the wire in opposite senses, in step 4, the two waveforms should be mirror images of each other as long as the ESD is not resulting in common mode interference to the measurement. This is another form of a null experiment. In step 5, the current probes are producing in-phase outputs and since they are close together on the wire, their outputs viewed on the scope will be almost identical. That is, about a 4 ampere peak current with a risetime faster than 1 nanosecond and a lower frequency ringing afterwards.

9.1.4   What effect does the metal plate have in step 6? Why?

Answer: The second current probe, on the downstream side of the metal plate, will show a much longer risetime, typically 10 to 30 nanoseconds while the upstream current probe will still show a risetime of less than 1 nanosecond. The lower frequency ringing will be nearly the same as displayed on the scope for both probes. This is because the metal plate has a much lower impedance to high frequency currents, even though it is floating, than does the grounded wire. This experiment shows that a piece of metal is a good sink for high frequency currents even if it is not grounded.

9.1.5   What happens to the waveforms if the ESD contact discharge is injected into the metal plate instead of at the end of the wire?

Answer: The two current probes will show inverted polarity of current rise from zero with respect to each other because as the current leaves the plate onto the wire, it flows through the current probes in opposite directions. The shape of the two current waveforms may not be the same because the impedances looking each way may not be equal. By paying attention to the phase difference between two current probes, much information about the location and nature of the source of the currents can be learned. This is similar to the two loop technique discussed in Chapter 7.

Experiment 9.2: Coaxial Current Probe ESD Event Logging

1. Connect a coaxial current probe, as shown in Figure 9.25, constructed of a 6 foot length of 50 ohm coaxial cable to a 50 ohm input of a digital oscilloscope. Connect a 6 foot wire to the probe tip and lay out the probe and wire straight on a table. Set the scope for 100 mV per division on the vertical scale, a trigger level of 50 mV positive slope, and a horizontal scale of 20 nanoseconds per division.
2. Generate an ESD spark either from an ESD simulator, or by walking across a rug and touching a metal key to a metal object, about 6 feet away from the coaxial current probe. Try several voltage levels if an ESD simulator is used and compare the displayed waveforms.

Questions:

9.2.1 How much current does the probe output indicate on the scope?

Answer: From several hundred milliamperes to as much as an ampere will be evident on the scope (remember that the transfer impedance of the probe is 1 ohm). In general, higher energy levels of ESD will generate higher currents in this test setup. By adjusting the scope's vertical sensitivity and trigger level, one can roughly estimate how many ESD events of a certain level are occurring in the area. There are many variables determining the actual current that is induced into the test setup, and many of those variables are difficult to control, so only a rough estimate is possible.

Index

Index